Polymer surfaces

Cambridge Solid State Science Series

EDITORS:
Professor R. W. Cahn
Applied Sciences Laboratory, University of Sussex
Professor M. W. Thompson
School of Mathematical and Physical Sciences, University of Sussex
Professor I. M. Ward
Department of Physics, University of Leeds

B. W. CHERRY

Associate Professor of Materials Engineering, Monash University

Polymer surfaces

CAMBRIDGE UNIVERSITY PRESS

Cambridge

London   New York   New Rochelle

Melbourne   Sydney

Published by the Press Syndicate of the University of Cambridge
The Pitt Building, Trumpington Street, Cambridge CB2 1RP
32 East 57th Street, New York, NY 10022, USA
296 Beaconsfield Parade, Middle Park, Melbourne 3206, Australia

First published 1981

Typeset by ⒸH Charlesworth & Co Ltd, Huddersfield

Printed in Great Britain at the
University Press, Cambridge

British Library Cataloguing in Publication Data

Cherry, B W
Polymer surfaces.   (Cambridge solid state science series).

1. Polymers and polymerization – Surfaces
I. Title   II. Series
547'.84   QD381   80-40013

ISBN 0 521 23082 9 hard covers
ISBN 0 521 29792 3 paperback

Contents

Preface

As polymers come to be used in ever more critical engineering applications, it becomes more important to understand the properties of their surfaces. Mechanical stress cannot be put into a material other than by means of stress transfer through its surface, and thus efficient utilisation of the properties of a material demands an understanding of the interactions which may take place at an interface. This is the prime reason for this study of polymer surfaces.

Historically, phenomena associated with the surfaces of polymers have been studied in a range of disciplines from physical chemistry to mechanical engineering, often from different viewpoints and usually with different terminologies, so that a chemist and an engineer working on the same system may have considerable difficulty communicating with each other. One object of the present book is therefore to try and present a logical development of the subject which spans most of the subject. Thus the book starts with a look at molecular interactions and surface thermodynamics, and then goes on to see how these control the chemistry of wetting and hence a major portion of adhesive technology. The strength of adhesive joints is controlled by the fracture mechanics of interfacial systems and the production of surfaces by rupture, and so an attempt is made to show how the surface chemistry of an adhesive system may influence the fracture mechanics of the system. Since some frictional phenomena involve the same molecular forces as adhesion, two chapters on friction attempt to show how the description of surface interactions developed in the earlier chapters explains these frictional phenomena. Wear is allied to friction, and the book concludes with a brief introductory treatment of this expanding subject.

A preface is the traditional place in which an author acknowledges the contribution made by others to the production of his work. Any book on this subject must depend heavily on the earlier books by Kaelble, by Wake and by Bowden & Tabor. I hope that I have given due credit where credit is due. Various colleagues have been prevailed upon to read and comment on various chapters, Bill Wake and John Griffiths have been responsible for major improvements in the text, and Ian Ward's reading of the whole book has gone far beyond the line of duty of a series editor. Thanks are due to the ladies who typed the script, Mrs Guthrie, Mrs Palmer and Mrs Fry, and to Julie Frazer who did a magnificent job photographing diagrams. To my wife and family, thanks are due for their forbearance during the protracted production of the book, and to Fricka, apologies for much neglect during the last two years.

B. W. Cherry

Melbourne
December 1979

1 Polymer surfaces

1.1 Introduction

This book is concerned with surfaces and with interfaces. The term interface will be used whenever we are concerned with phenomena which result from the interaction of substances which are on each side of the interface; the term surface will be used whenever we are concerned with phenomena resulting from the interactions of material which is only on one side of the interface. This chapter will be concerned with polymer surfaces: in fact these surfaces will always be in contact with a vapour phase but, because the concentration in the vapour phase is so low, in this chapter the presence of the vapour phase will be neglected.

The objective of this chapter is to define the thermodynamic variables which characterise the properties of a surface, to indicate how these properties may be determined experimentally, and then to relate those thermodynamic variables to the molecular properties of the material, both for solid and liquid surfaces. The most useful parameters for the discussion of surfaces are the Helmholtz free energy and the entropy, and so most of the discussion will be concerned with these quantities.

1.2 Basic thermodynamics of surfaces
Surface variables

Those atoms or molecules of a material which are situated close to its surface are subjected to different intermolecular forces from those to which molecules in the bulk of the material are subjected. Consequently the total energy of the system differs from the value which it would have had in the absence of the surface. It seems logical therefore to define a surface variable as the excess of that variable associated with the system due to the presence of the surface. The surface internal energy E^σ is thus given by the expression

$$E^\sigma = E - E^\beta, \tag{1.1}$$

where E is the total internal energy of the system and E^β is the total internal energy which the system would have had if all the material had been present in the unperturbed bulk state. In the same way it is possible to define a surface entropy (S^σ) and a surface Helmholtz free energy (A^σ) by means of the expressions

$$S^\sigma = S - S^\beta,$$
$$A^\sigma = A - A^\beta.$$

In general the Helmholtz free energy is a more useful variable in surface studies than the more commonly used Gibbs free energy, because nearly all surface studies are carried out at constant volume. A surface by geometrical definition has location and area but no volume, and so changes in the total amount of surface present can take place at constant volume.

Surface quantities are, of course, associated with a certain area of surface, Ω. The thermodynamic quantity per unit area is then given a lower case symbol and often termed a 'specific' surface variable. Thus the specific surface internal energy $e^\sigma = E^\sigma/\Omega$ and the specific surface entropy $s^\sigma = S^\sigma/\Omega$. The specific surface Helmholtz free energy is sometimes loosely termed the surface energy and given the symbol $a^\sigma = A^\sigma/\Omega$.

Surface tension and surface energy

A system possesses excess surface energy because the molecules in the surface are subjected to a different environment from those in the bulk. Because the molecules in the surface, unlike the molecules in the bulk, are subject to intermolecular attractions from one side only, the packing at least will differ from surface to bulk. If we now consider a line in the surface of the body and imagine the body divided into two by means of a plane perpendicular to the surface and containing the line, then similar arguments can be applied to the forces parallel to the surface across the dividing plane. Because of the difference in the molecular packing between the surface and the bulk, there will be a difference in the force acting across the dividing plane. The excess force per unit length of the line in the surface is termed the surface tension and is given the symbol γ; it is reckoned positive if it acts in such a direction as to contract the surface.

The surface tension may be related to the surface energy by considering the changes in the thermodynamic quantities which accompany a small change in a surface-containing system. Considering first a system which does not contain a surface, if N_i is the number of moles of the ith component which are present in the system, the chemical potential μ_i may be written as $\mu_i = (\partial A/\partial N_i)_{V,T,N_j}$. Consequently if dq is a small heat input to the system

$$dE = dq - d\omega + \sum_i \mu_i\, dN_i, \qquad (1.2)$$

where $d\omega$ is the work done by the system.

However, if the system contains a surface, work can be done on the system by extending the surface against surface tension forces, i.e.

$$d\omega = PdV - \gamma d\Omega.$$

Hence writing TdS for dq

$$dE = TdS - PdV + \gamma d\Omega + \sum_i \mu_i\, dN_i. \qquad (1.3)$$

Now by definition

$$A = E - TS$$

so that

$$dA = dE - TdS - SdT.$$

Hence, substituting from (1.3)

$$dA = \gamma d\Omega - PdV - SdT + \sum_i \mu_i \, dN_i$$

or

$$\gamma = \left(\frac{\partial A}{\partial \Omega}\right)_{V,T,N_i}. \tag{1.4}$$

The surface tension (a force per unit length) is thus equal to the change in Helmholtz free energy of the *whole system* associated with unit increase of surface area (an energy per unit area). It is not necessarily equal to the surface energy, which is the change in Helmholtz free energy of the *surface* associated with unit increase of surface area. The relationship between the surface tension and the surface energy may be derived as follows. The force necessary to extend the surface of an isotropic solid is γ per unit length, and so the work done in extending a surface by $d\Omega$ is $\gamma d\Omega$. The work done must equal the increase in total surface energy dA^σ. Therefore,

$$\gamma d\Omega = dA^\sigma = d(\Omega a^\sigma),$$

i.e.

$$\gamma = \frac{d}{d\Omega} (\Omega a^\sigma)$$

$$\gamma = a^\sigma + \Omega \left(\frac{da^\sigma}{d\Omega}\right). \tag{1.5}$$

For a liquid, any attempt to extend the surface will usually result in more molecules flowing into the surface, whose composition is thereby unchanged, i.e. $(da^\sigma/d\Omega) = 0$. For a liquid, therefore, the surface tension γ equals the surface energy a^σ. For a solid, however, as the surface is stretched the surface density of molecules changes and so $(da^\sigma/d\Omega) \neq 0$, and in general for a solid the surface tension and the surface energy are different.

Surface entropy and surface internal energy

The general expression

$$\left(\frac{\partial A}{\partial T}\right)_V = -S$$

may be written for a surface in the form

$$\frac{\partial}{\partial T}(\Omega\gamma) = -S^\sigma$$

or

$$-S^\sigma = \Omega \left(\frac{\partial \gamma}{\partial T}\right)_V + \gamma\left(\frac{\partial \Omega}{\partial T}\right)_V. \tag{1.6}$$

In general the coefficient of thermal expansion of a surface is small compared with the thermal coefficient of the free energy, and so the second term on the right-hand side of equation (1.6) may be neglected. Consequently the specific surface entropy is given by the expression

$$s^\sigma = -\left(\frac{\partial \gamma}{\partial T}\right)_V, \tag{1.7}$$

and the specific surface internal energy by the expression

$$e^\sigma = \gamma - T\left(\frac{\partial \gamma}{\partial T}\right)_V. \tag{1.8}$$

1.3 Experimental methods for the determination of polymer surface energies

Many of the methods available for the determination of the surface tension of low molecular weight liquids can be adapted for the determination of the surface tension and hence the surface energy of liquid polymers. The determination of the surface tension of solid polymers will be dealt with in section 1.5. In this section techniques which have been used for liquid polymers will be discussed and the techniques will arbitrarily be divided into those which involve the determination of the shape of a polymer surface and those which involve a determination of the load on a polymer surface.

The shapes of a liquid polymer surface

The shape of any liquid surface is governed (Aveyard & Haydon, 1973, p. 59) by Laplace's capillary equation

$$\Delta P = \gamma \left(\frac{1}{r_1} + \frac{1}{r_2}\right), \tag{1.9}$$

where ΔP is the pressure difference across a curved surface and r_1 and r_2 are the principal radii of curvature of the surface. So determination of the relationship between r_1, r_2 and ΔP therefore simply yields a value for the surface tension. Methods based on this technique include the sessile drop or bubble, the pendant drop, the maximum bubble pressure and the capillary rise.

The sessile bubble technique was used by Sakai (1965) for the determination of the surface tension of polyethylene melts. If a bubble in a liquid is trapped beneath a horizontal surface, as in figure 1.1, then if the bubble is large enough the lower surface becomes planar and horizontal.

Equation (1.9) can then be applied to either of the curved surfaces in the section through the drop shown in figure 1.1. The radius of curvature r_1 in the plane perpendicular to the paper is so much larger than the radius of curvature in the plane of the paper that $1/r_1$ can be neglected by comparison with $1/r_2$. Hence, since $\Delta P = (\rho_1 - \rho_v)gz$, where ρ_1 and ρ_v are the densities of the liquid and the vapour respectively and z is the distance above the planar polymer surface at 0,

$$\frac{\gamma}{r_2} = (\rho_1 - \rho_v)gz. \tag{1.10}$$

The radius of curvature of any curve is given by

$$r = \frac{[1 + (dz/dx)^2]^{3/2}}{d^2z/dx^2},$$

so that writing q for dz/dx and substituting in (1.10) yields

$$\frac{\gamma q\,dq}{(1 + q^2)^{3/2}} = (\rho_1 - \rho_v)gz\,dz,$$

which can be integrated using the fact that at $z = h, q = \infty$ to give

$$\frac{\gamma}{(1 + q^2)^{1/2}} = \frac{(\rho_1 - \rho_v)g\,(h^2 - z^2)}{2}.$$

Then since $q = 0$ when $z = 0$ at the bottom horizontal surface and ρ_v can usually be neglected by comparison with ρ_1,

$$\gamma = \tfrac{1}{2}\rho_1 g\,h^2,$$

and a determination of the depth of the drop below the equatorial plane *AB* will suffice to determine the surface tension and the surface energy of the liquid. Correction factors to modify the simple theory detailed above have been tabulated by Porter (1933) and Staicopolus (1962). Precisely the same theory applies to a drop of molten polymer on a flat plate (Cherry, el Mudarris & Holmes, 1969).

The basic theory of the pendant drop technique for the determination of the surface tension of a liquid was formulated by Andreas, Hauser & Tucker (1938), and adapted to systems which would solidify and retain their shape on cooling to room temperature by Davis & Bartell (1948). This method involves determining the shape of a liquid drop which hangs from the bottom of a vertical tube. If d_e is the equatorial diameter of a pendant drop, then by a similar operation to that which led to the equation for a sessile bubble, the relationship

Fig. 1.1. The sessile bubble.

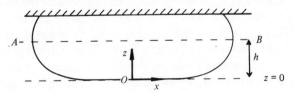

$$\gamma = (\rho_1 - \rho_v) g \, d_e^2 / H$$

may be derived, where H is a correction factor which varies with the shape of the drop and values for which have been tabulated by Adamson (1967). This method has been used by Wu (1969) and Roe (1968) for the determination of the surface tension of a range of polymer melts.

In the maximum bubble pressure method, bubbles are formed at a fine tip under the surface of the liquid. The radius of curvature of the bubble is a minimum, and hence the pressure in the bubble is a maximum when the bubble forms a hemisphere on the tip as diameter. Under these circumstances equation (1.9) leads to the result

$$\Delta P = 2\gamma/r_c,$$

where r_c is the radius of the capillary. However, the pressure across the surface will vary with the depth below the surface of the liquid and so the bubble will not be entirely hemispherical at the point of detachment. This leads to a small correction factor being applied to give (Edwards, 1968), for a tip which is situated at the surface of the fluid,

$$\gamma = \tfrac{1}{2}\Delta P r_c \left[1 - 2a(\rho_v - \rho_1)g/3\Delta P \right].$$

With viscous polymeric liquids the pressure which is measured will, unless the rate of application of the pressure is infinitely slow, include a component which is utilised in overcoming the forces of viscous deformation. In order to overcome this, the technique which has to be adopted is that bubbles are blown at successively lower pressures, with correspondingly longer periods to the bubble breaking away from the tip of the capillary. It is then possible to find the pressure at which the bubble will just not separate from the tip. Using this technique Edwards (1968) was able to determine the surface tension of a range of liquid polyisobutylenes.

The application of equation (1.9) to the rise of a liquid polymer in a capillary tube leads, assuming a zero contact angle for the polymer on the surface of the tube and neglecting any deviations from sphericity in the meniscus, to the expression

$$\gamma = \tfrac{1}{2}(\rho_1 - \rho_v) \, h r_c g,$$

where h is the rise in the capillary tube.

Schonhorn, Ryan & Sharpe (1966) developed a technique based on the determination of the capillary rise for their measurement of the surface tension of a liquid polychlorotrifluorethylene. Instead of the more usual technique of using a cathetometer to determine the difference in heights of the capillary rise in tubes of different radii, Schonhorn *et al.* located the position of the meniscus by probing with a very finely pointed wire attached to a micrometer screw. The instant of contact could be very easily seen and this technique was claimed by the authors to provide an accurate value for the height of the liquid in the capillary and to eliminate sources of error involved in the use of two capillary tubes.

Determination of surface tension by force measurements

Although a determination of the force required to stretch a polymer surface would seem an attractive proposition as a method of determining the surface tension, problems arise because during the stretching process energy has also to be expended in bringing about viscous deformation of the liquid, and whilst this second mechanism of energy dissipation is negligible for lower molecular weight liquids it may be considerable in the case of highly viscous polymer liquids. Two methods have, however, been developed which overcome this problem by using very carefully controlled rates of deformation; they are modifications of the du Nuouy ring technique and the Wilhelmy plate technique, which have been used for less viscous liquids. The du Nuouy ring technique, in which the force required to detach a horizontal ring from a liquid surface is measured, was used by Schonhorn & Sharpe (1965) to determine the surface tension of molten polyethylene. The elementary theory of the method suggests that just before detachment of the ring from the surface, the weight of liquid which is lifted out of the surface is entirely supported by the surface tension of the liquid acting vertically on either side of the ring. The mass of liquid W which is supported by the wire ring just before detachment should therefore be given by

$$Wg = 4\pi a_r \gamma, \tag{1.11}$$

where a_r is the radius of the ring. Schonhorn & Sharpe used an Instron tensile testing machine in their work. By utilising the moving cross head of the machine to withdraw the ring at a series of standard rates, and the load cell to measure the force required for detachment, they could, by extrapolation, determine the detachment force at zero rate of withdrawal when the viscous forces would be non-existent. The simple theory represented by equation (1.11) is seriously in error due to curvature of the polymer surfaces, but correction factors calculated using equation (1.9) have been published by Harkins & Jordan (1930).

The modification to the Wilhelmy plate technique which was used by Dettre & Johnson (1966) to determine the surface tension of molten polyethylene is illustrated in figure 1.2.

If the tip of the plate just touches the surface of the molten polymer, then for zero contact angle the force due to surface tension is $p\gamma$, where p is the perimeter of the plate. Then if ΔWg is the difference between the weight of the plate in air and its weight at equilibrium in contact with the surface but at zero immersion,

$$\Delta W g = p\gamma + W_1 g,$$

where $W_1 g$ is the weight of the thin film of liquid which creeps up above the meniscus. Dettre & Johnson showed that at a constant rate of immersion W_1 was a linear function of the depth of prior immersion and so were able to calculate W_1 and hence γ for the polymer melt.

Experimental values for the surface tension of liquid polymers

Wu (1974) has written an excellent review of the techniques available for the determination of surface tensions, and reports the values given in table 1.1 for surface tensions at 180 °C.

Although it may be noted that the polymers with the larger surface energies in table 1.1 are those in which dipole–dipole interactions play a part in the molecular bonding, it can also be seen that the increase in surface energy above that which may be ascribed solely to dispersion force interactions is small. This point will be returned to later when it will be

Fig. 1.2. Modified Wilhelmy plate technique for the determination of the surface tension of molten polymers. (From Dettre & Johnson, 1966, p. 369.)

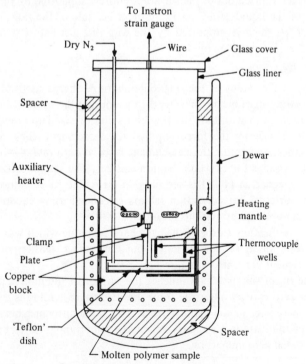

Table 1.1. *Typical values for surface tension of some liquid polymers at 180 °C*

Polymer	γ (mN m^{-1})	$-\partial\gamma/\partial T$ (mN m^{-1} °C^{-1})	Method	Reference
Polytetrafluorethylene	9.4	0.062	Wilhelmy plate	[a]
Polypropylene (atactic)	20.4	0.056	Pendant drop	[b]
Polyethylene (linear)	26.5	0.057	Pendant drop	[c]
Polymethyl methacrylate	28.9	0.076	Pendant drop	[d]
Poly(α-methyl styrene)	29.4	0.058	Sessile bubble	[e]
Polyethylene terephthalate	(33.6)*	–	Rotating bubble	[e]
Polyamide mixture	(35.6)*	–	Rotating bubble	[e]

*These are calculated values; the measured values at 290 °C are 27.0 and 29.0 respectively, and a value of $(\partial\gamma/\partial T)$ of -0.06 has been assumed.
[a]Dettre & Johnson (1969). [d]Wu (1974).
[b]Roe (1968). [e]Wu (1970).
[c]Wu (1969).

seen that dispersion force interactions may be responsible for the major portion of adhesive forces between phases.

1.4 Calculation of the surface energy

Since surface phenomena all stem from the fact that the inter-molecular forces acting on the surface molecules are acting only on one side of the surface, methods for the calculation of the surface energy on theoretical grounds must either seek to correlate surface properties with those bulk properties which are controlled by the same intermolecular forces, or must seek to calculate the surface properties from a knowledge of the intermolecular potential functions.

Surface tension and cohesive energy density

The internal energy of a polymer must have, in addition to a component which is due to short-range nearest neighbour interactions, a component which is due to the long-range structure of the molecule. This second component will take account of such diverse effects as chain entanglements and cross-linking and long-range polar effects. The molar internal energy $\bar{E}$ may therefore be written as

$$\bar{E} = \bar{E}_n + \bar{E}_d, \tag{1.12}$$

where $\bar{E}_n$ is the contribution due to short-range intermolecular forces and $\bar{E}_d$ is the contribution due to the long-range structure. If $\bar{V}$ is the molar volume, then $\bar{E}/\bar{V}$ is termed the cohesive energy density and given the symbol δ^2 where δ is termed the solubility parameter. $\bar{E}_d$ is independent of the molar volume whereas $\bar{E}_n$ is dependent upon the molar volume. Equation (1.12) can be rewritten

$$\bar{E} = \tfrac{1}{2} N_A z \epsilon + \bar{E}_d,$$

where ϵ is the energy of attraction between two molecules, z is the average number of nearest neighbours of a molecular unit, N_A is Avogadro's number and the factor $\tfrac{1}{2}$ occurs since each molecule is counted twice in the summation.

Since the total internal energy is a function of both z and V,

$$d\bar{E} = \left(\frac{\partial \bar{E}}{\partial \bar{V}}\right)_z d\bar{V} + \left(\frac{\partial \bar{E}}{\partial z}\right)_{\bar{V}} dz. \tag{1.13}$$

The 'physical state parameter', n, is defined (Kaelble, 1971, p. 57) by the expression

$$n = -\left(\frac{\partial E}{\partial V}\right)\bigg/\frac{E}{V},$$

so that

$$\left(\frac{\partial \bar{E}}{\partial \bar{V}}\right)_z = -n\left(\frac{\bar{E}}{\bar{V}}\right) = -\frac{n(\bar{E}_n + \bar{E}_d)}{\bar{V}}.$$

Now

$$\bar{E}_n = \tfrac{1}{2} N_A z \epsilon, \tag{1.14}$$

therefore

$$\left(\frac{\partial \bar{E}_n}{\partial z}\right)_{\bar{V}} = \frac{N_A \epsilon}{2} = \frac{\bar{E}_n}{z},$$

so that (1.13) may be rewritten

$$d\bar{E} = \bar{E}_n \frac{dz}{z} - n(\bar{E}_n + \bar{E}_d)\frac{d\bar{V}}{\bar{V}}. \tag{1.15}$$

Now if $d\bar{E}^\sigma$ is the change in the internal energy of the system when one mole of material is moved from the bulk of a phase to the surface

$$d\bar{E}^\sigma = \bar{E}_n \left(\frac{z^\sigma - z}{z}\right) - n(\bar{E}_n + \bar{E}_d)\left(\frac{\bar{V}^\sigma - \bar{V}}{\bar{V}}\right).$$

If the surface area occupied by one mole of the surface species is Ω_{N_A}, then since the area occupied by one molecule is $(\bar{V}/N_A)^{2/3}$,

$$\Omega_{N_A} = N_A (\bar{V}/N_A)^{2/3} = N_A^{1/3} \bar{V}^{2/3}.$$

Now

$$e^\sigma = d\bar{E}^\sigma/\Omega_{N_A},$$

therefore

$$e^\sigma = \frac{\bar{V}^{1/3}}{N_A^{1/3}} \left[\frac{\bar{E}_n}{\bar{V}}\left(\frac{z^\sigma - z}{z}\right) - n\frac{\bar{E}}{\bar{V}}\left(\frac{\bar{V}^\sigma - \bar{V}}{\bar{V}}\right)\right]. \tag{1.16a}$$

In order to clarify the nature of the second term in equation (1.16a) it is necessary to invoke the concept of the 'internal pressure' (Hildebrand & Scott, 1950, p. 99). P_i is defined as $(\partial E/\partial V)_T$. For any system the external pressure exerted by that system is given by

$$P = -\left(\frac{\partial A}{\partial V}\right)_T = -\left(\frac{\partial E}{\partial V}\right)_T + T\left(\frac{\partial S}{\partial V}\right)_T,$$

but

$$\left(\frac{\partial S}{\partial V}\right)_T = \left(\frac{\partial P}{\partial T}\right)_V,$$

therefore

$$P = -\left(\frac{\partial E}{\partial V}\right)_T + T\left(\frac{\partial P}{\partial T}\right)_V.$$

Thus the observed pressure is the difference between the internal pressure $P_i = (\partial E/\partial V)_T$, due to attractive forces between molecules, and the kinetic pressure P_k due to the repulsive forces. Under normal conditions P is negligible compared with the other two terms so that

$$\left(\frac{\partial E}{\partial V}\right)_T \approx T\left(\frac{\partial P}{\partial T}\right)_V = T\left(\frac{\partial V}{\partial T}\right)_V \Big/ \left(\frac{\partial V}{\partial P}\right)_V,$$

i.e.

$$\left(\frac{\partial E}{\partial V}\right)_T \approx -T\frac{\alpha}{\beta},$$

where $\alpha = (1/V)(\partial V/\partial T)$ is the coefficient of thermal expansion and $\beta = (1/V)(\partial V/\partial P)$ is the compressibility. Now from the previous definition of the 'physical state parameter', n,

$$n\left(\frac{E}{V}\right) = -\frac{\partial E}{\partial V} \approx T\frac{\alpha}{\beta},$$

and so substituting in equation (1.16a)

$$e^\sigma = \left(\frac{\bar V}{N_A}\right)^{1/3}\left[\frac{\bar E_n}{\bar V}\frac{(z^\sigma - z)}{z} + T\frac{\alpha}{\beta}\frac{(\bar V^\sigma - \bar V)}{\bar V}\right]. \tag{1.16b}$$

For a liquid, since $a^\sigma = \gamma$,

$$e^\sigma = \gamma - Ts^\sigma,$$

and so comparing this with (1.16b)

$$\gamma = \left(\frac{\bar V}{N_A}\right)^{1/3}\frac{\bar E_n}{\bar V}\frac{(z^\sigma - z)}{z}, \tag{1.17a}$$

and

$$s^\sigma = \left(\frac{\bar V}{N_A}\right)^{1/3}\left[\left(\frac{\alpha}{\beta}\right)\frac{(\bar V - \bar V^\sigma)}{\bar V}\right]. \tag{1.17b}$$

Equation (1.17a) represents a theoretical justification for the empirical relationship reported by Hildebrand & Scott (1950, p. 402)

$$(\gamma/\bar{V}^{1/3}) = \text{const } (\bar{E}/\bar{V}),$$

where $(\bar{E}/\bar{V})$ was determined from the heat of vaporisation. Schonhorn (1965) and Wu (1968) have subsequently developed this expression further in order to account for the effect of chain length upon surface energy.

Surface tension and the parachor

McLeod (1923) reported that the surface tension of a liquid could be related to the difference in density between the liquid and its vapour by the relation

$$\gamma = C (\rho_1 - \rho_v)^m, \tag{1.18}$$

where C and m are constants, and for low molecular weight liquids $m \approx 4$.

Roe (1968) showed that equation (1.18) adequately represented the variation in surface tension with temperature for a number of polymers, but that m was less than 4, though usually greater than 3, and Wu (1969, 1974) used McLeod's equation to develop expressions for the molecular weight dependence of the surface tension, and to relate the surface tension to the glass transition temperature.

Sugden defined the parachor (Quayle, 1953) by the relation

$$\tilde{P} = C^{1/4} M,$$

where M is the molecular weight, so that, if the vapour density can be neglected in comparison with the liquid density,

$$\gamma = \tilde{P}^4 \, \rho_1^4/M^4. \tag{1.19}$$

The parachor for a given liquid is the sum of the parachors of the constituent parts of the molecule, and Roe (1965b, 1968) has shown that the surface tension of a polymer liquid may be predicted from equation (1.19) if values for the repeat unit for $\tilde{P}$ and M are utilised.

Surface tension and molecular potential functions

A more formal approach to the calculation of the surface energy of a liquid from a knowledge of the intermolecular interaction potentials would be to form the partition function for the system and to derive the surface energy from the partition function by means of the relation

$$A = -kT \ln Z,$$

where Z is the partition function and k is Boltzmann's constant. The calculation of the partition function is beyond the scope of this chapter. The technique has however been well described by Defay, Prigogine, Bellemans & Everett (1966) and has been applied to the calculation of the surface energy of polyethylene by Roe (1965a) and by Stewart & Von Frankenberg (1968).

1.5   The measurement of the surface energy of solid polymers

Two methods have been proposed for the determination of the surface energy of a solid by direct measurement. These involve either the stretching of thin threads, or extrapolation from the liquid state. A third method based on the direct measurement of the force between two solids will be described later.

The stretching of thin polymeric filaments under their own weight

When a thin filament of a solid hangs vertically it is subjected to two opposing influences; under the action of its own weight it tends to increase in length, whereas the action of the surface forces will be such as to cause retraction. If the material is capable of undergoing plastic deformation under the action of the forces involved then the surface energy may be determined from the balance between the two forces.

If a thin filament hangs vertically it will neither extend nor retract when the energy utilised in the creation of new surface area equals the work done on the system by gravitational forces, i.e. when

$$\gamma d\Omega = Wg dh, \tag{1.20}$$

where Wg is the weight of the filament and h is the height of the centre of gravity above some reference point. If the radius of the filament is r and its length is l, then since $dh = \frac{1}{2}dl$,

$$2\gamma d(2\pi rl) = Wg \, dl. \tag{1.21}$$

Since the deformation of the system is purely plastic, it takes place at constant volume, and so writing $\pi r^2 l$ as V_f and introducing the density of the material ρ, equation (1.21) becomes

$$2\gamma\pi^{1/2} V_f^{1/2} l^{-1/2} \, dl = \rho V_f g \, dl,$$

or

$$\gamma = \tfrac{1}{2} \rho grl_0, \tag{1.22}$$

where l_0 is the length of a filament which neither extends nor contracts.

Greenhill & McDonald (1953) used this technique to determine the surface free energy of solid paraffin wax by suspending filaments of wax in a constant temperature enclosure and plotting $\Delta l/l\Delta t$ (positive or negative) against t and interpolating the value of l_0 at which $\Delta l/l\Delta t$ equals zero. Phillips & Riddiford (1966) suggested however that the values obtained (60–70 mJ m^{-2} between 30 and 50 °C) were probably too high as the mobility of the paraffin molecules would be too low to allow them to come to equilibrium in their lowest free energy state.

The extrapolation of solid surface energies from the liquid region

For an amorphous solid there should be no discontinuous change in the free energy of either the bulk or the surface phases as the material cools down from the fluid state, and Fowkes & Sawyer (1952) have demonstrated by means of contact angle measurements (which will be dealt with in chapter 2) that, for a fluorinated lubricating oil which was a waxy solid at room temperature, calculations based on this premise may be justified by experiment. It should therefore be possible to measure the surface tension of the material over a range of temperatures in the liquid state, and then to estimate the surface tension of the solid by extrapolation to lower temperatures.

If a phase change occurs during the cooling from liquid temperatures, then from McLeod's equation (1.18), a discontinuous change in density will lead to a discontinuous change in the surface tension. If, for example, crystallisation takes place and ρ_c and ρ_a are the densities of the crystalline and amorphous phases respectively, then from equation (1.18) it can be seen (Wu, 1974) that

$$\gamma^c = (\rho_c/\rho_a)^m \, \gamma^a,$$

where γ^c and γ^a are the surface tensions of the crystalline and amorphous regions respectively.

1.6 Calculation of the theoretical surface energy of a solid

The calculation of the surface energy of a solid must follow the same approach as has already been utilised for a liquid. However, since the low strain mechanical properties of a solid are related to the intermolecular forces, it is possible to relate surface properties to the mechanical properties for solids, and so calculations involving more detailed understanding of the intermolecular forces are included here. The following argument is based on the concept of a plane which divides a column of material of unit cross-sectional area into two pieces, as in figure 1.3. If the nature of the interactions across the plane is understood then a calculation of the work which must be expended against intermolecular forces to separate the two pieces of material will lead to a value of twice the surface energy (Good, 1967; Gardon, 1967); alternatively a summation of all the interactions across the plane will lead to an expression for the surface internal energy (Fowkes, 1969). The surface energy will be calculated below on the basis of an approach put forward by Fowler & Guggenheim (1939) who state that the formulae derived 'are essentially those of Rayleigh (1890) in modern dress'.

If the energy of interaction of two molecules is given by the Lennard-Jones expression

$$\epsilon = -\frac{A}{r^6} + \frac{B}{r^{12}}, \tag{1.23}$$

where r is the distance of separation of the molecules, then the force between them is given by

$$F = \frac{\partial \epsilon}{\partial r} = \frac{6A}{r^7} - \frac{12B}{r^{13}}.$$

If the density of the molecules, n_1, is independent of their position, then the total number in the shaded annulus in figure 1.3 is given by M, where

$$M = 2\pi r \sin \theta \ (dr/\sin \theta) \ n_1 \ df.$$

The force in the z-direction exerted on a molecule located at X due to all molecules in the annulus is given by $F_a = MF \cos \theta = MF(f/r)$, and so the force due to all the molecules in the slab of width df is given by F_s where

$$F_s = 2\pi n_1 \ fdf \int\limits_{r=f}^{r=\infty} \left(\frac{6A}{r^7} - \frac{12B}{r^{13}} \right) \ dr.$$

Consequently the force due to all the molecules in the left-hand block acting on the molecule at X is given by $F_b = \int_{f=j}^{f=\infty} F_s df$, and the total force between all the molecules in the left-hand block and the right-hand block is given by F_t, where

$$F_t = \int\limits_{j=a}^{j=\infty} F_b n_1 \ dj$$

and a is the distance of separation of the two blocks, i.e.

$$F_t = 2\pi n_1^2 \int\limits_{j=a}^{j=\infty} dj \int\limits_{f=j}^{f=\infty} fdf \int\limits_{r=f}^{r=\infty} \left(\frac{6A}{r^7} - \frac{12B}{r^{13}} \right) \ dr = \frac{2\pi n_1^2}{a^3} \left(\frac{A}{12} - \frac{B}{90a^6} \right).$$

$$(1.24)$$

Fig. 1.3. Intermolecular forces across an interface.

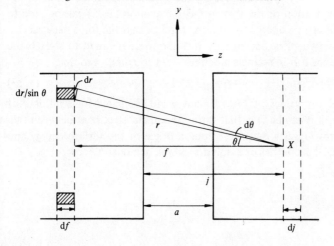

If two surfaces are formed by separating the right- and left-hand sides of the material shown in figure 1.3 then the work done per unit area of the original cross-section should be equal to twice the specific surface energy since two surfaces are formed. The surface energy should therefore be given by

$$2\gamma = \int_{a=r_{11}}^{a=\infty} F_t \, da,$$

where r_{11} is the equilibrium distance between the planes. The subscript '11' is introduced to indicate that the planes under consideration each form the surface of a phase of type 1. Therefore,

$$\gamma = \frac{\pi n_1^2}{24 r_{11}^2} \left(A - \frac{B}{30 r_{11}^6} \right), \tag{1.25}$$

However, if r_{11} is the equilibrium distance between the two semi-infinite bodies, the net force between them should be zero. Therefore, from equation (1.24), for the condition $a = r_{11}$

$$\frac{A}{12} = \frac{B}{90 r_{11}^6},$$

whence, substituting in equation (1.25)

$$\gamma = \pi n_1^2 A / 32 \, r_{11}^2. \tag{1.26}$$

Equation (1.26) is not, in fact, strictly accurate since it was derived on the assumption that the structure remains constant right up to the surface, whereas in reality the surface structure differs from the bulk. It is also inaccurate because the derivation neglects changes in structure (e.g. Poisson's ratio effects) occasioned by the action of the force which brings about the separation of the two surfaces. Equation (1.26) can be used to calculate the surface energy by utilising the fact that the force necessary to separate the two planes is initially just the force required to stretch the material. Hence if E is Young's modulus, $E = (\partial F_t r / \partial r)_{r_{11}}$, so that

$$E = \pi n_1^2 A / r_{11}^3, \qquad \gamma = E \, r_{11}/32. \tag{1.27}$$

In table 1.2 the values of $E r_{11}/32$ and γ are compared for a number of polymers, and it can be seen that the approximate agreement between two sets of figures suggests that the common origin of the surface energy and the mechanical properties of a polymer is well demonstrated.

Table 1.2. *Inter-relationship of surface energy and modulus for some polymers*

Polymer	Elastic modulus ($N\ m^{-2}$)	r_{11} (nm)	Surface energy ($mJ\ m^{-2}$) at 25 °C	
			Calculated	Observed
Polytetrafluoroethylene	4.5×10^{8}	0.48	6.8	25
Polyethylene	11.0×10^{8}	0.43	15.0	30
Polymethyl methacrylate	36.0×10^{8}	0.52	58.0	40

2   The solid–liquid interface

2.1   Introduction

Chapter 1 was primarily concerned with surfaces, surface energy and surface tension, and it entirely neglected the gaseous phase with which the solid or liquid surface must have been in contact; chapter 2 is, however, concerned with interfaces, interfacial energy and interfacial tension, i.e. it is concerned with those situations in which material on both sides of the interface plays a role in determining the properties associated with the interface.

The nomenclature for this chapter differs slightly from that adopted in chapter 1, since the symbol for an interfacial property must indicate the nature of the phases on either side of the interface; thus a_{sv}^σ refers to the interfacial energy of the solid-vapour interface, and where it is necessary to consider the hypothetical case of a solid in contact with a vacuum the symbol a_s^σ will be used, as in chapter 1. A similar convention will be used for surface tensions.

Whether a solid-liquid interface is formed by the growth of a crystal within a supercooled melt, or by the spreading of a liquid over a surface, the molecules at the interface are subjected to intermolecular forces directed towards the bulk of each phase. In this chapter, therefore, an attempt will be made to show how the properties of the interface are controlled by the bulk properties of both phases, and then how the physical formation of the interface is controlled by the properties of the interface.

2.2   The interfacial tension and the work of adhesion

Surface tensions were shown in chapter 1 to be scalar quantities which may, therefore, be added algebraically. Hence, if two surfaces could be placed in contact without any intermolecular interaction the interfacial tension, γ_{12}, would be given by

$$\gamma_{12} = \gamma_1 + \gamma_2.$$

There must, however, be intermolecular interactions, and so the 'work of adhesion', W_a, may be defined as the decrease in Helmholtz free energy of the whole system due to these interactions, i.e.

$$W_a = \gamma_1 + \gamma_2 - \gamma_{12}. \tag{2.1a}$$

The work of adhesion, therefore, also represents the work necessary to separate two surfaces which meet at an interface. The 'work of cohesion',

W_c, may similarly be defined as the decrease in Helmholtz free energy when two similar surfaces coalesce to give a homogeneous body, i.e.

$$(W_c)_1 = 2\gamma_1 \text{ and } (W_c)_2 = 2\gamma_2. \tag{2.1b}$$

The work of cohesion, therefore, also represents the reversible work necessary to create new surfaces by cleaving a homogeneous body. From equation (1.26) it can be seen that

$$(W_c)_1 = \pi n_1^2 A_{11}/16r_{11}^2 .$$

In precisely the same way as was used for equation (1.26), it is possible to calculate the work of adhesion between two phases, and the analogue of equation (1.26) is

$$(W_a)_{12} = \pi n_1 n_2 A_{12}/16r_{12}^2. \tag{2.2}$$

Girifalco & Good (1957) discussed the ratio of the work of adhesion of an interface to the geometric mean of the individual works of cohesion. They gave this ratio the symbol ϕ, so that

$$\phi = \frac{(W_a)_{12}}{[(W_c)_{11}(W_c)_{22}]^{1/2}} = \frac{A_{12}}{(A_{11}A_{22})^{1/2}} \frac{r_{11}r_{22}}{(r_{12})^2} . \tag{2.3}$$

Good (1967) has reviewed some of the arguments for the relationships between the attractive constants, A_{11}, A_{22} and A_{12}, and has concluded that if the molecules are interacting solely by means of dispersion forces then the relationship may be written

$$A_{12} = (A_{11}A_{22})^{1/2}. \tag{2.4}$$

This is the so-called 'geometric mean' relationship, and if it is incorporated in (2.3) we obtain

$$\phi = r_{11}r_{22}/r_{12}^2 .$$

Alternatively, writing the works of adhesion and cohesion in terms of the individual surface and interfacial tensions,

$$\phi = \frac{\gamma_1 + \gamma_2 - \gamma_{12}}{2(\gamma_1\gamma_2)^{1/2}} ,$$

or

$$\gamma_{12} = \gamma_1 + \gamma_2 - 2\phi(\gamma_1\gamma_2)^{1/2} \tag{2.5a}$$

and

$$W_a = 2\phi(\gamma_1\gamma_2)^{1/2}. \tag{2.5b}$$

The validity of equations (2.5) does not of course depend upon the applicability of the geometric mean relationship of equation (2.4). However, for systems in which dispersion forces predominate and for which the geometric mean relationship for the attractive constants holds, then it is found that ϕ is often within 20% of unity, so that the approximation

$$\gamma_{12} = \gamma_1 + \gamma_2 - 2(\gamma_1 \gamma_2)^{1/2},$$

may be used.

Interfacial energies and mixed molecular interactions

The value of the constant ϕ in equations (2.5) only reduces to unity when dispersion force interactions predominate on both sides and across the interface. In order to develop the appropriate relationship between the interfacial tension and the surface tensions of the phases on either side of the interface a little more generally, Fowkes (1969) suggested that the work of adhesion between two phases across an interface is given by a simple summation of the works of adhesion due to each specific type of intermolecular interaction, e.g.

$$W_a = W_a^d + W_a^h + W_a^p + W_a^i + W_a^\pi + W_a^{da} + W_a^e + \ldots, \tag{2.6}$$

where superscripts refer to, London dispersion forces, hydrogen bonds, dipole–dipole interactions, dipole–induced-dipole interactions, π-bonds, donor–acceptor bonds, and electrostatic interactions. Since the work of cohesion can be written in a similar form to equation (2.6), and since $(W_c)_1 = 2\gamma_1$, the surface tension may also be written as the sum of its component parts in the form

$$\gamma_1 = \gamma_1^d + \gamma_1^h \ldots.$$

Fowkes also suggested that the only significant interactions which occur across an interface between two different phases are those which are due to those types of interaction which are common to both phases. Dispersion forces are common to all materials, and so W_a^d is often a dominant term on the right-hand side of equation (2.6), and when one of the phases has only dispersion force interactions, W_a can be essentially equal to W_a^d.

It has been seen above that for systems in which the interfacial interactions stem from dispersion forces, ϕ in equation (2.5b) is approximately one. Hence, the dispersion force component of the work of adhesion may be written

$$W_a^d = 2(\gamma_1^d \gamma_2^d)^{1/2}. \tag{2.7}$$

Consequently, if an interface separates two phases which have only dispersion force interactions in common, then $W_a = W_a^d$ and the interfacial tension will be given by (from equations (2.7) and (2.1a))

$$\gamma_{12} = \gamma_1 + \gamma_2 - 2(\gamma_1^d \gamma_2^d)^{1/2}, \tag{2.8}$$

All the treatments which lead to a relationship between the interfacial tension and the surface tensions of the individual phases on either side of the interface have been based on the geometric mean relationship

$$A_{12} = (A_{11} A_{22})^{1/2} \tag{2.4}$$

for the attractive constants. Wu (1973) has suggested that it may be equally valid to consider the harmonic mean, defined by

$$A_{12}^h = \frac{1}{2}\left(\frac{1}{A_{11}} + \frac{1}{A_{22}}\right) \tag{2.9}$$

as representing the interfacial attractive constant. Since the conclusions which are reached based on the use of the harmonic mean attractive constant do not differ markedly from those reached using a geometric mean, and since the calculations based on the geometric mean are usually simpler, the geometric mean will be used throughout the rest of this chapter.

2.3 The crystal-melt interface

When the melt of a semi-crystalline polymer is cooled, at some temperature below the equilibrium melting point crystals of solid polymer will start to grow. The growth of the crystalline phase is controlled by the balance between the loss in free energy of the system due to the change of a certain mass of material from the liquid to the solid state and the gain in free energy due to the creation of an area of solid-liquid interface. At the equilibrium melting point of the crystal, T_f, the change in free energy must be zero, and so since

$$\Delta A = \Delta E_f - T_f \Delta S_f = 0,$$

where ΔE_f and ΔS_f are the changes in internal energy and entropy on fusion, respectively: $\Delta S_f = \Delta E_f/T_f$. Now the change in entropy on crystallisation is likely to be independent of the temperature at which crystallisation takes place, and so remembering that $(\partial\Delta A/\partial T)_V = -\Delta S$, the change in free energy on crystallisation at a temperature T lower than T_f will be given by $\Delta A'$, where

$$\Delta A' = 0 - \frac{\partial\Delta A}{\partial T}(T_f - T) = \frac{T_f - T}{T_f}\Delta E_f. \tag{2.10}$$

Since this is the free energy which must be utilised in the creation of the crystal-melt interface, it can be seen that the magnitude of the interfacial energy controls crystal growth. Schonhorn (1965) has suggested that the actual interfacial energy of a semi-crystalline solid in contact with its melt is a function of the degree of crystallinity of the solid, and has proposed an equation, analogous to (1.17a), for such a system

$$a_{sl}^\sigma = (1 - x_2)(\bar{V}_s/N_A)^{1/3}(\Delta\bar{H}_f/\Delta\bar{V}_f)K_s,$$

where the subscript sl refers to the solid-liquid interface, x_2 is the degree of crystallinity for the solid phase, $\bar{V}_s$ is the molar volume of the solid phase, $\Delta\bar{H}_f$ and $\Delta\bar{V}_f$ are the molar heat of fusion and change in molar volume on fusion, respectively, and K_s is a 'binding constant'. Making the appropriate substitutions, a_{sl}^σ for polyethylene varies between 0 for a completely amorphous solid to 117 mJ m^{-2} for a completely crystalline polymer.

Experimental determination of the crystal-melt interfacial energy

The determination by Brown & Eby (1964) of the interfacial energy of the polyethylene crystal-melt interface will serve both as an illustration of the role played by the interfacial energy in controlling crystal growth and an example of the determination of the interfacial energy.

Polyethylene crystallises in folded chain lamellae. If a growing fibril is as illustrated in figure 2.1, and another layer of molecules is added to the end of the fibril, this changes the total area of solid-liquid interface, and so the change in free energy of the whole system is given by $\Delta A''$:

$$\Delta A'' = 2bl\gamma_{side} + 2ab\gamma_{end} + \Delta a'abl, \qquad (2.11)$$

where l is the thickness of the lamella, a the width of the lamella and b the height of the additional layer of molecules as shown in figure 2.1. γ_{side} and γ_{end} are the interfacial energies of the sides and ends of the folded loops in contact with the melt, respectively, and $\Delta a'$ is the free energy per

Fig. 2.1. Growth of a chain folded lamella within a spherulite.

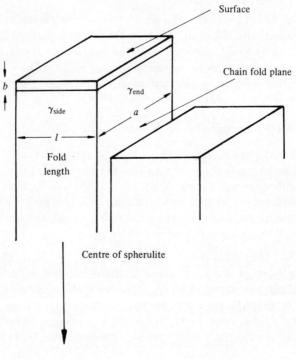

unit volume liberated by a molecule on going from the liquid to the crystalline state. From equation (2.10), $\Delta a'$ can be written as

$$\Delta a' = \frac{T_f - T}{T_f} \, \Delta e_f,$$

and so, substituting in (2.11)

$$\Delta A'' = 2bl\gamma_{side} + 2ab\,\gamma_{end} + abl\,\Delta e_f(T_f - T)/T_f.$$

A folded chain crystal will melt at a temperature below the equilibrium melting point of the extended chain crystal. This will occur at the temperature for which $\Delta A'' = 0$, i.e. when

$$2\gamma_{side}/a + 2\gamma_{end}/l = -\Delta e_f(T_f - T)/T_f. \tag{2.12}$$

In general $a \gg l$, so that the first term on the left-hand side of (2.12) can be neglected, and so

$$T = T_f(1 - 2\gamma_{end}/l\Delta e_f). \tag{2.13}$$

Equation (2.13) can be utilised (Gornick & Hoffman, 1966) to obtain both T_f and γ_{end}. To determine l either low-angle X-ray diffraction or electron microscopy may be used; a plot of T against l^{-1} yields an intercept at $l^{-1} = 0$ of T_f, and the slope is $-2\gamma_{end}T_f/\Delta e_f$. Using this method, Brown & Eby (1964) obtained a value of $\gamma = 57 \pm 5$ mJ m^{-2} for polyethylene, which is consistent with the calculation of Schonhorn cited earlier.

2.4   Wetting and contact angles

The alternative method by which a solid–liquid interface may be formed is for a liquid to spread over a solid surface. The thermodynamic driving force for the formation of an interface is (considering only surface forces) given by

$$(\mathrm{d}w)_V = -\mathrm{d}A = -\sum_i \gamma_i \, \mathrm{d}\Omega_i. \tag{2.14}$$

It should be noted, however, that the earlier restriction that the liquid is well below its critical temperature and hence has a negligible vapour pressure no longer applies. Consequently, the vapour of the liquid phase will be adsorbed on the solid surface, reducing its surface energy and surface tension. The amount of the reduction in surface tension is termed the spreading pressure, π_s, so that

$$\pi_s = \gamma_s - \gamma_{sv}, \tag{2.15}$$

where the subscript sv in equation (2.15), and subsequently, indicates the solid surface in equilibrium contact with the vapour of the spreading liquid, and the subscript s indicates the hypothetical case of a solid in contact with a vacuum.

Equation (2.14) can now be written out in full, for the formation of unit area of solid-liquid interface, in order to define the 'adhesion tension', T, as the thermodynamic force which promotes wetting

$$T = \gamma_{sv} - \gamma_{sl}. \tag{2.16}$$

The adhesion tension is, in fact, a much more useful concept than γ_{sv} or γ_{sl} since it is extremely difficult to measure γ_{sv} or γ_{sl} but quite easy to measure their difference by reference to the contact angle.

Contact angles

When a drop of liquid rests on a surface, unless spreading takes place the liquid-vapour interface will form a finite angle with the (assumed planar) solid surface. This angle is termed the contact angle and usually given the symbol θ. The relationship between the contact angle and the properties of the liquid and solid has been a matter of controversy for many years (Johnson, 1959), but the following derivation is similar to the classical derivation propounded by Gibbs, although in the form presented below it neglects the effects of adsorption, gravity or second-order terms in the surface areas.

If the edge of a drop is as illustrated in figure 2.2, then the system will be in equilibrium when for a virtual displacement from the equilibrium position $\delta A = 0$. Now if the symbol $(\delta A)_{lv}$ is used to denote the change in Helmholtz free energy of the whole system due to a change in the liquid-vapour interfacial area, then

$$(\delta A)_{lv} = -dw = \gamma_{lv} d\Omega_{lv}.$$

Hence, by a simple extension of the notation,

$$\delta A = (\delta A)_{sv} + (\delta A)_{sl} + (\delta A)_{lv}.$$

If the edge of the drop (the base of which is assumed to be a circle of radius R) advances from B to C so that the periphery of the drop is now AC, whereas before it was AB, then the change in free energy is given by

$$\delta A = -2\pi R dR \, \gamma_{sv} + 2\pi R dR \, \gamma_{sl} + 2\pi R dS \, \gamma_{lv}.$$

Fig. 2.2. Contact angle between a drop of liquid and the polymer surface on which it rests. (From Cherry, 1971, p. 222.)

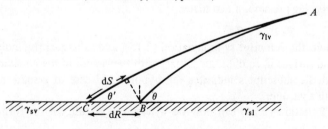

But $dS = dR \cos \theta'$, and so

$$\delta A = 2\pi R dR \left(-\gamma_{sv} + \gamma_{sl} + \gamma_{lv} \cos \theta'\right),$$

and since at equilibrium $\delta A = 0$, $\theta' = \theta$

$$\gamma_{sv} - \gamma_{sl} = \gamma_{lv} \cos \theta. \tag{2.17}$$

Equation (2.17) is termed the Young-Dupré equation. It has been criticised, e.g. by Bikerman (Johnson, 1959), on the grounds that the vertical component of the liquid-vapour surface tension, i.e. γ_{lv}, $\sin \theta$ has no balancing force. However, it has been shown (Lester, 1967) that the component $\gamma_{lv} \sin \theta$ does deform the surface, but that this does not destroy the validity of the Young-Dupré equation when the deformation can be neglected, which is usually the case.

The combination of equations (2.16) and (2.17) yields

$$T = \gamma_{lv} \cos \theta, \tag{2.18}$$

and so the adhesion tension is now expressed in easily determinable quantities.

Contact angles and dispersion force contributions to the surface tension

Equations (2.15) and (2.17) can be combined with equation (2.8), written for a solid-liquid interface as

$$\gamma_{sl} = \gamma_s + \gamma_l - 2(\gamma_s^d \gamma_l^d)^{1/2}, \tag{2.19}$$

to give the expression

$$\cos \theta = \frac{2(\gamma_s^d \gamma_l^d)^{1/2}}{\gamma_{lv}} - \frac{\gamma_l}{\gamma_{lv}} - \frac{\pi_s}{\gamma_{lv}}, \tag{2.20}$$

Now, since the adsorption of its own vapour is unlikely to modify the surface energy of a liquid, $\gamma_{lv} = \gamma_l$, and for systems for which the liquid does not spread spontaneously over the surface Fowkes (1964) has suggested that π_s may be negligible. Hence equation (2.20) reduces to

$$\cos \theta = -1 + \frac{2(\gamma_s^d \gamma_l^d)^{1/2}}{\gamma_l}, \tag{2.21}$$

and so if the contact angle for a number of liquids for which $\gamma_l = \gamma_l^d$ (typically hydrocarbon liquids and similar) is measured on a single polymer surface, then the result of a plot of $\cos \theta$ against $\gamma_l^{-1/2}$ should be a straight line which intersects the ordinate at -1 and which has a slope of $2(\gamma_s^d)^{1/2}$. The line intersects the $\cos \theta = +1$ line when $\gamma_s^d = \gamma_l$. Such a graph (Fowkes, 1967) is shown in figure 2.3.

The critical surface tension for wetting

Zisman (1964) made a different approach to the wetting of polymer surfaces by low molecular weight liquids. The ability of a drop of a

Fig. 2.3. Variation of cos θ with surface tension of the liquid – equation (2.21). The solids are: A polyethylene, B paraffin wax, C $C_{36}H_{74}$, D a perfluorodecanoic acid monolayer on platinum. (From Cherry, 1971, p. 229.)

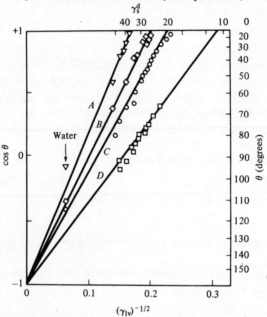

liquid to spread over a solid is governed by the spreading coefficient S, (not to be confused with the spreading pressure π_s) which is the decrease in free energy of the system when a drop spreads over a surface. In this case the terms in equation (2.14) must include one representing the upper surface of the drop, and so

$$S = \gamma_{sv} - \gamma_{sl} - \gamma_{lv}, \tag{2.22}$$

and unless S is positive the liquid will not spread spontaneously in the absence of external forces. For metals and other inorganic high melting point solids, γ_{sv} is usually between 0.5 and 1.5 J m^{-2} and most liquids spread over them easily. Soft organic compounds, on the other hand, have much lower surface energies – they are often known as low surface energy solids – and S can only remain positive if γ_{lv} is small.

Zisman investigated how the contact angle formed by various liquids on smooth low-energy surfaces varied with the surface tension of the liquid. His results are illustrated in figure 2.4. It may be seen that the results may be represented by the empirical relationship

$$\cos \theta = b - c\, \gamma_{lv}, \tag{2.23}$$

where b and c are constants for a homologous series of liquids on a given polymer surface. The apparent discrepancy between equation (2.23) and equation (2.21) will be returned to later.

Fig. 2.4. Contact angle on various perfluorinated low-energy surfaces by *n*-alkane liquids. (From Cherry, 1971, p. 231.)

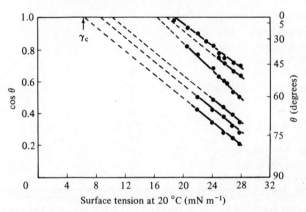

As γ_{lv} in equation (2.23) decreases, θ decreases until at a value of γ_{lv} = $(b - 1)/c$ the liquid spreads over the polymer surface and the contact angle is zero. This value of the surface tension of the liquid above which spontaneous spreading does not take place is termed the critical surface tension, is given the symbol γ_c and is found to be a characteristic of the surface of the solid.

The combination of the Young-Dupré equation and the definition of the spreading coefficient yields

$$S = \gamma_{lv} (\cos \theta - 1).$$

It can, therefore, be seen that surface forces only promote the spreading process (i.e. S is not negative) when the surface tension of the liquid is less than the critical surface tension of the solid. Of course this does not imply that if $\gamma_{lv} > \gamma_c$ and the spreading coefficient is negative that spreading is impossible; gravitational and other forces may cause the liquid to spread over the surface.

Molecular constitution and the critical surface tension

The critical surface tensions for some typical polymers are shown in table 2.1 (Johnson & Dettre, 1969).

Zisman explained the order in which the polymers appear in table 2.1 in a qualitative fashion by suggesting that the dispersion force interactions are less for fluorine-containing groups than other hydrocarbons, and that the force fields-associated with -CH_3 or -CF_3 groups are less than those associated with the corresponding -CH_2- or -CF_2- groups. Increasing wettability then follows the increasing availability of interactions other than pure dispersion forces.

Table 2.1. *Critical surface tensions for some typical polymers (at 20 °C)*

Polymer	γ_c (mN m^{-1})
Polytetrafluoroethylene	18
Polytrifluoroethylene	22
Polyvinylidene fluoride	25
Polyvinyl fluoride	28
Polyethylene	31
Polytrifluorochloroethylene	31
Polystyrene	33
Polyvinyl alcohol	37
Polyvinyl chloride	39
Polyvinylidene chloride	40
Polyethylene terephthalate	43
Polyhexamethylene adipamide	46

It can be seen from table 2.1, however, that the trend for the critical surface tension for wetting, γ_c, to vary with molecular constitution follows that of the variation of the surface tension of the solid, γ_s. Good (1973) attempted to clarify the relationship between γ_c and γ_s by considering the effect of interactions across the solid–liquid interface due to factors other than dispersion forces.

If interactions occur by polar forces as well as dispersion forces, equation (2.19) might be written in the form (Owens & Wendt, 1969)

$$\gamma_{sl} = \gamma_s + \gamma_l - 2(\gamma_s^d \gamma_l^d)^{1/2} - 2(\gamma_s^p \gamma_l^p)^{1/2}, \qquad (2.24)$$

or in the more general form, since $\gamma_s = \gamma_s^d + \gamma_s^p$, and $\gamma_l = \gamma_l^d + \gamma_l^p$

$$\gamma_{sl} = [(\gamma_s^d)^{1/2} - (\gamma_l^d)^{1/2}]^2 + [(\gamma_s^p)^{1/2} - (\gamma_l^p)^{1/2}]^2.$$

The introduction of (2.24) into the argument that gave (2.21) leads to

$$\cos \theta = \frac{2(\gamma_l^d \gamma_s^d)^{1/2}}{\gamma_l} + \frac{2(\gamma_l^p \gamma_s^p)^{1/2}}{\gamma_l} - 1, \qquad (2.24')$$

and so if the interaction of a liquid with a non-polar solid is considered, since $\gamma_s^p = 0$ and $\gamma_s = \gamma_s^d$,

$$\cos \theta = -1 + \frac{2(\gamma_s \gamma_l^d)^{1/2}}{\gamma_l},$$

but for a liquid which just wets the surface $\gamma_l = \gamma_c$ and $\cos \theta = 1$, i.e.

$$\gamma_c = (\gamma_l^d \gamma_s)^{1/2}. \qquad (2.25)$$

Two possible situations now arise. If the liquid is non-polar $\gamma_l^d = \gamma_l = \gamma_c$ and $\gamma_c = \gamma_s$. In other words, the critical surface tension which is found by the use of a series of liquids which have no polar interactions will be equal to the surface tension of the solid, if it likewise is completely non-polar. If the liquid is capable of polar interactions, however, $\gamma_l = \gamma_l^d + \gamma_l^p$, and substitution in (2.25) and solution of the resulting quadratic expression for γ_c yields the expression

$$\gamma_c = \tfrac{1}{2}\gamma_s [1 \pm (1 - 4\,\gamma_l^p/\gamma_s)^{1/2}].$$

It is already known that if $\gamma_l^p = 0$ then $\gamma_c = \gamma_s$, and so the positive root must be taken leading to

$$\gamma_c/\gamma_s = \tfrac{1}{2}[1 + (1 - 4\,\gamma_l^p/\gamma_s)^{1/2}],$$

from which it can be seen that the presence of polar intermolecular attractions in the liquid can reduce γ_c below γ_s to a minimum value of $\tfrac{1}{2}\gamma_s$.

The dispersion force component and hence the polar force component of a liquid such as water can be determined by measuring the interfacial tension between water and a liquid in which only dispersion forces occur. Using equation (2.8), in which γ_2^d is the only unknown, the dispersion force component for water is found to be 21.8 mJ m^{-2}. From figure 2.3, γ_s^d for paraffin wax can be seen to be 25 mJ m^{-2}. From equation (2.21) the contact angle for water on paraffin wax should be 111°. The measured value for the contact angle of water on paraffin wax (Fowkes & Harkins, 1940) is found to be $110 \pm 2°$.

A comparison of approaches to the calculations of contact angles

Equations (2.21) and (2.23) would appear to be incompatible since the former predicts a dependence of $\cos\theta$ on $(\gamma_{lv})^{-1/2}$, whereas the latter predicts a dependence of $\cos\theta$ on $(-\gamma_{lv})$. Phillips & Riddiford (1966) have, however, pointed out that both treatments may be equivalent if π_s for these systems is not zero, as postulated by Fowkes, but a specific function of γ_{lv} and γ_{ls}. They have suggested that if π_s can be expressed as a power series in $(\gamma_s\gamma_l)^{1/2}$ then both treatments are compatible.

2.5 Wetting of rough and inhomogeneous surfaces

All the discussion of section 2.4 was concerned with the spreading of a liquid on a smooth flat surface. However, in practice such surfaces rarely occur, and the effect of surface roughness must now be examined.

If the surface is rough (but not so rough as to cause the formation of voids at the solid–liquid interface) and the roughness factor, that is, the ratio of true surface area to apparent surface area is r then, for this surface, equation (2.16) must be modified to give

$$T^R = r(\gamma_{sv} - \gamma_{sl}),$$

where T^R is the adhesion tension of the rough surface. From equation (2.18), if ϕ is the apparent contact angle on the rough surface

$$T^R = \gamma_{lv} \cos\phi, \tag{2.26}$$

and so since from equation (2.17)

$$\gamma_{sv} - \gamma_{sl} = \gamma_{lv} \cos \theta,$$

it can be seen that

$$\cos \phi / \cos \theta = r. \tag{2.27}$$

This expression is known as Wenzel's equation (Wenzel, 1936) and from it, it can be seen that if the intrinsic contact angle for a liquid on a solid surface is less than $90°$, roughening the surface will reduce the contact angle and promote wetting.

Cassie & Baxter (1944) extended Wenzel's treatment to the case in which a composite interface is formed, that is when air is trapped in the 'valleys' in the solid surface, as shown in figure 2.5. In this case, in unit apparent area of the composite interface, the solid-liquid interface may have an area Ω_{sl}, and the liquid-vapour interface an area Ω_{lv}. By the same reasoning that led to equation (2.16) the adhesion tension T^C for a composite interface is given by

$$T^C = \Omega_{sl}(\gamma_{sv} - \gamma_{sl}) - \Omega_{lv} \gamma_{lv},$$

and since for the apparent contact angle ϕ, $T^C = \gamma_{lv} \cos \phi$, the apparent contact angle for the composite surface is given by

$$\cos \phi = \Omega_{sl} \cos \theta - \Omega_{lv}. \tag{2.28}$$

Cassie & Baxter pointed out that equation (2.28) is a special case of the general equation for the apparent contact angle at a composite interface for which the intrinsic contact angles were θ_1 and θ_2, which is

$$\cos \phi = \Omega_1 \cos \theta_1 + \Omega_2 \cos \theta_2.$$

In the case of equation (2.28), θ_2 is $180°$.

The formation of a composite interface at high values of surface roughness explains the water repellency of some fabrics and materials such as ducks' feathers. The intrinsic contact angle for water on the material of a duck's feather is only $95°$, but because the feathers are covered in fine hairs, each only about 8 μm in diameter and held apart at a distance of

Fig. 2.5. Formation of a composite interface. (From Cherry, 1971, p. 236.)

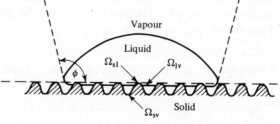

about 150°, water droplets just roll off. Hence the expression 'like water off a duck's back'!

Contact-angle hysteresis

The treatment so far has been concerned with equilibrium values of the contact angle, and of the wetting. These equilibrium values are not, however, achieved instantaneously when a drop of liquid is placed on a polymer surface, and the kinetics of the approach of a drop of liquid to its equilibrium shape must now be studied. A study of the kinetics of wetting will also highlight another factor which has previously been neglected – contact-angle hysteresis: this is the observation that contact angles measured at the edge of a drop which is advancing over a rough or heterogeneous surface may be greater than the values measured when the liquid is receding – usually by only a small amount, but sometimes by as much as 80°.

An explanation of contact-angle hysteresis which links it to surface roughness has been put forward by Johnson & Dettre (1969). They considered a drop spreading over a rough surface for which the intrinsic contact angle was θ and the apparent contact angle was ϕ. If, as is shown in figure 2.6, the angle of the rough surface to the horizontal at any point is α, then in any 'valley' there are only two configurations which are possible, that is when the drop edge is at such a position that $\phi = \theta - \alpha$, where α can be positive or negative. If a simple profile is assumed for the surface, then the value of T^R or T^C can be calculated for each possible configuration of the drop. Johnson & Dettre carried out such a calculation and found that the absolute free energy minima corresponded to the value which would be predicted by the equations of Wenzel or Cassie & Baxter, respectively, depending upon whether a composite or non-composite surface was chosen as a basis for the calculation of the free energies. For the

Fig. 2.6. Effect of a composite interface on the apparent contact angle. (From Cherry, 1971, p. 236.)

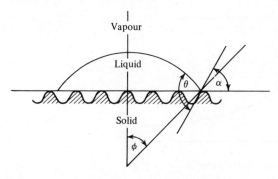

two possible configurations allowed to a drop edge in each valley, Johnson & Dettre found that each corresponded to a local maximum or minimum value of the free energy, and that the configuration in which the edge of the drop is close to the top of the trough represents a local minimum, and the configuration in which the edge of the trough is close to the bottom represents a local maximum.

Thus as a drop spreads over a surface it encounters a series of metastable configurations separated by activation energy barriers. The free energies of the metastable configurations decrease towards the equilibrium (Wenzel or Cassie & Baxter) configuration, but the activation energy barriers are found to increase towards the equilibrium configuration. The approach of a drop to its equilibrium configuration will usually depend upon thermal activation to move the advancing drop over the activation energy barriers, and when the height of the barriers increases, the rate of approach to equilibrium will be reduced.

Accordingly, Johnson & Dettre suggest that the observed hysteresis in contact-angle measurements results from the fact that these angles are measured on drops which are only in metastable equilibrium. Figure 2.7 shows their results which have been calculated for a series of surfaces similar to that depicted in figure 2.6 but characterised by different roughness factors. The lines labelled *A* and *B* represent the maximum and

Fig. 2.7. Effect of roughness on contact angle for $\theta = 120°$: *A* maximum possible angles, *B* minimum possible angles, *C* possible curve of receding angles, *D* possible curve of advancing angles, *E* most probable contact angles calculated from Wenzel's equation, *F* most probable contact angles calculated from Cassie & Baxter's equation (From Cherry, 1971, p. 236.)

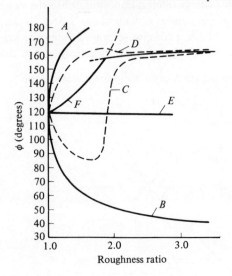

minimum values, respectively, of ϕ which are allowed for purely geometrical constraints. The lines D and C represent values which may be observed for advancing and receding contact angles if it is assumed that the edge of the drop is stopped at an activation energy barrier of a given height (Johnson & Dettre 1964), and the lines labelled E and F represent the equilibrium configurations corresponding to the equations of Wenzel and Cassie & Baxter, respectively. Calculations of the free energy show that as soon as a certain roughness is exceeded the formation of a composite interface is favoured and the hysteresis is reduced considerably.

3.1   The nature of adhesion

Chapter 2 was concerned with intermolecular interactions across an interface between a polymer and a second phase. These intermolecular attractions must be responsible for the adhesion between the two phases and so, as the first step in an attempt to determine what causes adhesive properties, the theoretical value of the force required to separate two phases which are perfectly in contact will be calculated.

In section 1.6 the total force between two semi-infinite blocks of the same material was calculated and the result obtained that

$$F_t = \frac{2\pi n_1^2}{a^3} \left(\frac{A}{12} - \frac{B}{90a^6} \right). \tag{1.24}$$

The same calculation could have been carried out with different materials on either side of the dividing interface and the result would have been

$$F_t = \frac{2\pi n_1 n_2}{a^3} \left(\frac{A_{12}}{12} - \frac{B_{12}}{90a^6} \right)$$

or, since when at r_{12} (the equilibrium distance between the phases) $F_t = 0$ and so $A_{12}/12 = B_{12}/90r_{12}^6$,

$$F_t = \frac{2\pi n_1 n_2 A_{12}}{12a^3} \left[1 - \left(\frac{r_{12}}{a} \right)^6 \right].$$

In order to separate the two phases, a force equal to the maximum value of F_t must be applied to the interface. This maximum value is given by the condition that $(\partial F_t/\partial a) = 0$, and hence

$$(F_t)_{max} = \frac{2\pi n_1 n_2 A_{12}}{12} \frac{2}{3^{3/2} r_{12}^3},$$

but by analogy with (1.26)

$$\gamma_{12} = \pi n_1 n_2 A_{12}/32r_{12}^2,$$

and so

$$(F_t)_{max} = 2.06 \gamma_{12}/r_{12}. \tag{3.1}$$

This represents the force necessary to separate two materials joined at an interface if, during separation, the surfaces remain parallel and only elastic work is done. Putting in realistic values, say $\gamma_{12} = 30$ mJ m^{-2}, $r_{12} = 3 \times 10^{-10}$ m, $(F_t)_{max} \approx 200$ MN m^{-2}.

This figure is of course way in excess of typical values for the strength of an adhesive joint, and the reasons are not hard to find. Normally the two surfaces will not remain parallel during the parting operation and there will be at least some degree of 'peel' involved, which allows the molecular interactions to be disrupted individually and sequentially. The interface is rarely if ever perfect in the sense that it is free from flaws, and defects at the interface give rise to stress concentrations so that the local stress may be many times the average applied stress. Even in the absence of an applied stress the interface in an adhesive joint may not be stress-free, as the shrinkage of the adhesive during setting, or differential thermal expansion and contraction of adhesive and adherend will give rise to self-equilibrating stresses within the joint system which may be great enough to disrupt the joint. Finally, adhesive joints are rarely if ever designed so that there is uniform stress transfer from adhesive to adherend, even in the absence of interfacial defects, and so again the stresses at part of the interface may be many times the mean applied stress.

However, the above calculation does point out two important aspects of adhesion studies. The first is that the above calculation was based on the order of magnitude of interfacial energy which might arise solely as a result of dispersion force interactions. If all other factors were absent, then extremely strong joints could be made which relied on dispersion force interactions and there would be no need to invoke any special 'forces of adhesion' to explain the strength of a joint. Obviously if there are other intermolecular interactions, or even if there is chemical bonding between the two phases, then they will enhance the strength of a joint, but such interactions are not a necessary condition of adhesion. The second important aspect is that since the theoretical strength of an adhesive joint is so high, the study of the factors which lead to the weakness of an adhesive joint are at least as important as the study of the strength. An important point which arises out of this is the difference between adhesion and joint strength. Joint strength is the physical force required to disrupt an adhesive joint of a given configuration when that force is applied in a specific manner. Adhesion, or more specifically the 'work of adhesion', defined in chapter 2 as

$$W_a = \gamma_1 + \gamma_2 - \gamma_{12},$$ (2.1a)

is a thermodynamic function which may be determined by one or more of the methods of physical chemistry. The relationship between the two is in general unknown even qualitatively, and only in the most specific cases is there any degree of quantitative understanding of the relationship between the two.

This section may, therefore, be summarised by the conclusion that an adhesive is any liquid which is capable of wetting a solid and subsequently solidifying in such a way that the shrinkage stresses induced by the change of state do not disrupt the joint. In this context it should be noted that water is an extremely good adhesive, but one which only functions below 0 °C, as the vast expenditure on de-icing equipment for aeroplane wings indicates.

3.2 Wetting and adhesion

One cause of weakness in an adhesive joint is the presence of defects at the adhesive–adherend interface, caused by the inability of the adhesive to penetrate all the crevices in the substrate surface completely. Micro-cracks at the interface act as stress concentrators which can greatly reduce the strength of the joint. It is therefore necessary to examine the factors which control the ability of a liquid to wet a surface.

The thermodynamics of wetting

The thermodynamic driving force for the formation of an interface is the spreading coefficient defined by equations (2.17) and (2.22) as

$$S = \gamma_{sv} - \gamma_{sl} - \gamma_{lv} = \gamma_{lv}(\cos\theta - 1).$$

So, for similar joints formed using a common adhesive on a substrate which has been prepared with a variety of surface treatments, the joint strength should vary in the same direction as $\cos\theta$. This is in fact the case, as can be seen from table 3.1 (Houwink & Salomon, 1965, p. 108) in which, although the correlation is not perfect, the adhesive joint strength can be seen to increase as the contact angle decreases.

The relationship between contact angle and joint strength also provides an explanation of the efficacy of the old craftsman's technique of 'roughing' a joint surface before applying glue, since an increase in surface roughness by Wenzel's equation (2.27) decreases the contact angle.

Table 3.1. *Correlation between contact angle of the adhesive and strength of the joint (steel adherend; epoxy resin adhesive; lap shear joint)*

Form of substrate	Contact angle (degrees)	Joint strength ($MN\ m^{-2}$)
As received	77	3.7
Washed in toluene	59	12.4
Washed in methyl ethyl ketone	47	12.6
Polished and washed	41	14.4
Grit blasted and washed	36	16.7
Sulphuric acid–dichromate etch	34	14.4
Hydrofluoric acid etch	29	15.4

The critical surface tension and adhesion

Henceforth the symbols γ_a and γ_s will be used to denote the surface energies of adhesive and substrate, respectively. Since the spreading coefficient is negative unless $\cos \theta$ is equal to 1 (and this will occur only when γ_a of the adhesive is less than the critical surface tension of the substrate $(\gamma_s)_c$), an adhesive will only function adequately when its surface tension is less than the critical surface tension of the substrate. Sharpe & Schonhorn (1964) demonstrated this point very elegantly by forming a strong joint when they used molten polyethylene ($\gamma_a \approx 31$ mJ m^{-2}) as the adhesive to bond substrates coated with an epoxy resin (($\gamma_s)_c \approx 45$ mJ m^{-2}), but when the situation was reversed and an attempt was made to bond solid polyethylene (($\gamma_s)_c \approx 31$ mJ m^{-2}) with a liquid epoxy resin ($\gamma_a \approx 45$ mJ m^{-2}), only very weak joints could be formed. Sharpe & Schonhorn suggested that this non-reciprocity of wetting offered an explanation of the adhesion rule propounded by de Bruyne many years earlier (de Bruyne, 1939). 'Provided we use pure or simple substances as adhesives then there is a good deal of evidence that strong joints can never be made by polar adherends with non-polar adhesives or to non-polar adherends with polar adhesives.' The first part of the rule is simply explained in that non-polar adhesives will normally have a critical surface tension which is well below the surface tension of a non-polar adhesive. De Bruyne is said (Sharpe & Schonhorn, 1964) to have suggested that whilst the second part of the rule covered all known examples when it was formulated, it could now be seen to be incorrect.

The variations in the surface treatments which have to be applied to various polymer surfaces before they can be used in the form of adhesive joints can now be reconciled with the critical surface tensions listed in table 2.1. The fluoro compounds are the best examples of 'non-stick' compounds, which can only be wetted with considerable difficulty. Fluoro polymers may be made wettable, and joints can be formed by treatments such as immersion in a solution of sodium in liquid ammonia, which presumably oxidises the surface and raises $(\gamma_s)_c$ to a higher value. The polyalkalines for which $(\gamma_s)_c$ is of the order of 30–35 mN m^{-1} require a less vigorous treatment in order to oxidise their surfaces; they are commonly treated by flaming, by subjection to a high-voltage spark discharge, or by immersion in concentrated oxidising acids. For polymers such as polyvinyl chloride or polyethylene terephthalate, for which $(\gamma_s)_c$ is close to 40 mN m^{-1}, it is found that a thorough cleaning of the surface is usually sufficient to promote good wetting.

Optimum wettability for adhesion

In chapter 2 it was pointed out that the Fowkes expression for interfacial energy could be modified to include interaction terms arising from polar components of the total intermolecular interactions. If the interfacial interactions can be expressed as a geometric mean term, then for the adhesive-substrate system

$$\gamma_{as} = \gamma_a + \gamma_s - 2(\gamma_a^d \gamma_s^d)^{1/2} - 2(\gamma_a^p \gamma_s^p)^{1/2}. \qquad (3.2)$$

The adhesion tension for this system is given as

$$T = \gamma_s - \gamma_{as},$$

and so substituting from (3.2),

$$T = 2(\gamma_a^d \gamma_s^d)^{1/2} + 2(\gamma_a^p \gamma_s^p)^{1/2} - \gamma_a.$$

For a given substrate, γ_s and γ_s^p are fixed. It is, however, possible to find the optimum polarity (defined as γ_a^p/γ_a) of an adhesive of given surface tension, γ_a, to maximise the wetting and the joint strength. This calculation was originally carried out by Wu (1973) for a system in which the intermolecular interactions were expressed by a harmonic mean equation, as in (2.9), but the same result is obtained with the geometric mean equation (3.2).

If the polarity of the substrate $(\gamma_s^p/\gamma_s) = \zeta$, and the polarity of the adhesive $(\gamma_a^p/\gamma_a) = \xi$, then

$$T = 2[(\gamma_a - \gamma_a^p)(\gamma_s - \gamma_s^p)]^{1/2} + 2(\gamma_a^p \gamma_s^p)^{1/2} - \gamma_a$$

$$= 2(\gamma_a \gamma_s)^{1/2} \{[(1 - \xi)(1 - \zeta)]^{1/2} + (\xi\zeta)^{1/2} - \tfrac{1}{2}(\gamma_a/\gamma_s)^{1/2}\}$$

For the maximum value of T, $(\partial T/\partial \xi)_{\gamma_a \gamma_s \zeta} = 0$, which leads to the result $\xi = \zeta$.

Thus the optimum thermodynamic wettability condition is obtained when the polarities of the adhesive and substrate are exactly the same. This is, of course, de Bruyne's rule in another guise.

The kinetics of wetting

The formation of an interface is controlled not only by the thermodynamics but also by the kinetics of wetting. This may be particularly important in the case of an industrial process for which it may not be possible to allow the necessary time for the system to come to thermodynamic equilibrium. Cherry (Cherry & Holmes, 1969; Cherry, el Mudarris & Holmes, 1969) examined the kinetics of wetting at the molecular level by adapting the treatment of Johnson & Dettre (1964) to account for the dependence of the rate of approach of a drop to its equilibrium configuration when the viscosity of the molten polymer which formed the drop, rather than the roughness of the substrate, was the major factor controlling the rate of approach to an equilibrium configuration.

If the activation energy for viscous flow for the polymer is ΔG_η, then if the decrease in free energy due to the action of surface forces is dw for forward movement of the drop edge and $-dw$ for backward movement of the drop edge, the rate constant for forward flow is given (Glasstone, Laidler & Eyring, 1941) as k', where

$$k' = (kT/h) \{\exp[-(\Delta G_\eta + \tfrac{1}{2}dw)/kT] - \exp[-(\Delta G_\eta - \tfrac{1}{2}dw)/kT]\}$$
$$= (2kT/h)\exp(-\Delta G_\eta/kT)\sinh(-dw/2kT),$$

or if $dw \ll 2kT$,

$$k' = (-dw/h)\exp(-\Delta G_\eta/kT), \tag{3.3}$$

where k is Boltzmann's constant, h is Planck's constant and T is the absolute temperature.

In order to calculate dw as the drop edge moves from a configuration represented by $\cos \phi_1$ to a configuration represented by $\cos \phi_2$, it will be assumed that the solid surface is flat and that the radius of the circle of contact, r, is much greater than the distance between successive points of metastable equilibrium. Hence, if a length y of drop edge advances a distance dr,

$$d\Omega_{sl} = y \, dr = -d\Omega_{sv}$$

and

$$d\Omega_{lv} = d\Omega_{sl} \cos \phi = y \, dr \cos \phi,$$

where the subscripts sl, sv and lv refer to the solid-liquid, solid-vapour and liquid-vapour interfaces, respectively. Hence dw is given by

$$dw = \int\limits_{\cos\phi_1}^{\cos\phi_2} (\gamma_{lv} \cos \phi + \gamma_{sl} - \gamma_{sv}) y \, dr,$$

or since

$$\gamma_{sl} - \gamma_{sv} = -\gamma_{lv} \cos \theta,$$

$$dw = \int\limits_{\cos\phi_1}^{\cos\phi_2} \gamma_{lv} y(\cos \phi - \cos \theta) \, dr. \tag{3.4}$$

Since the change in ϕ as the drop moves forward between two adjacent positions of metastable equilibrium is small, $\cos \phi$ may be considered to vary linearly with r between $\cos \phi_1$ and $\cos \phi_2$, i.e.

$$\cos \phi = \cos \phi_1 + c(\cos \phi_2 - \cos \phi_1),$$

where $c = (r - r_1)/(r_2 - r_1)$. Now if x is the distance travelled by the drop edge between free-energy maxima, $x = (r_2 - r_1)$ and $dr = x \, dc$, so that (3.4) may be rewritten as

$$dw = -\gamma_{lv} xy \int\limits_0^1 [\cos \theta - \cos \phi_1 - c(\cos \phi_2 - \cos \phi_1)] \, dc,$$

and on integrating, and remembering that $(\cos \phi - \cos \phi_1) \gg (\cos \phi_2 - \cos \phi_1)$,

$$\mathrm{d}w = -\gamma_{lv}\, xy\, (\cos \theta - \cos \phi).$$

So that substituting in equation (3.3) yields

$$k' = (\gamma_{lv}\, xy/h)\exp(-\Delta G_\eta/kT)(\cos \theta - \cos \phi). \tag{3.5}$$

The Eyring treatment of viscous flow (Glasstone *et al.*, 1941) yields for the viscosity (η) of a fluid

$$\eta = (h/v)\exp(\Delta G_\eta/kT),$$

where v is the activation volume (which may be written as $v = xyL$ where L is the length of the flow unit in the x-direction). Hence, substituting in equation (3.5), and putting $k' = \mathrm{d}\cos\phi/\mathrm{d}t$,

$$\frac{\mathrm{d}\cos \phi}{\mathrm{d}t} = \frac{\gamma_{lv}}{\eta L}(\cos \theta - \cos \phi). \tag{3.6}$$

Equation (3.6) is the same expression that had been observed empirically by Schonhorn, Frisch & Kwei (1966) and Newman (1968), and shows that as a drop approaches its equilibrium configuration its rate of approach to that configuration decreases to zero.

3.3   Rheology and adhesion

Equation (3.6) was, of course, developed for the case of an adhesive spreading across a flat horizontal surface, and the activation energy barriers were ascribed to the viscous processes of the liquid flow. In many techno-logical situations, however, the solid surface will be rough and the adhesive will have to penetrate comparatively deep crevices within the surface under the action of an applied pressure. Under such circumstances, Cassie & Baxter's equation (2.28) does not apply, since the driving force for the for-mation of the interface stems from the applied pressure, and not from interfacial forces. The rate at which a viscous adhesive can penetrate a crevice in a model surface can, however, be calculated, making several sim-plifying assumptions so that an estimate of the order of magnitude of the times involved may be obtained.

A model discussed by Bikermann (1968, p. 69) consists of a long V-shaped groove in the surface of the substrate, as shown in figure 3.1, in which the axis of the groove is supposed to be perpendicular to the paper. When the groove is partially filled with adhesive as shown, then the pressure across the curved surface is, by Laplace's capillary equation (1.9), given by

$$\Delta P = \gamma_{lv}/r$$

since the radius of curvature of the adhesive in the plane perpendicular to the paper is infinite. Since from figure 3.1 it can be seen that $x_1 = r\cos(\theta - \alpha)$

$$\Delta P = \gamma_{lv} \cos(\theta - \alpha)/x_1. \tag{3.7}$$

If this pressure acts on an element dx wide, as shown in the inset to figure 3.1, then if the adhesive behaves as a Newtonian liquid of viscosity η, and if the speed of vertical movement into the groove is u, then it can be seen that

$$\Delta P\, dx = \eta y\, \frac{\partial u}{\partial x} - \eta y \left(\frac{\partial u}{\partial x} - \frac{\partial^2 u}{\partial x^2} \right) dx \quad ,$$

or

$$\frac{\Delta P}{y} = \eta\, \frac{\partial^2 u}{\partial x^2},$$

which can be integrated, remembering that at $x = 0$, $\partial u/\partial x = 0$ and that at $x = x_1$, $u = 0$, to yield

$$\Delta P(x^2 - x_1^2) = 2y\eta u.$$

Now the mean velocity of the leading surface of the adhesive as it penetrates the groove is given by

$$\bar{u} = \frac{\Delta P}{2y\eta} \int\limits_0^{x_1} \frac{(x^2 - x_1^2)}{x_1} dx = -\frac{\Delta P}{3y\eta}\, x_1^2.$$

The negative sign arises because the centre of the adhesive surface is actually moving upwards relative to the edges. Now, writing $dy/dt = \bar{u}$, putting $x_1 = x_0(1 - y/y_0)$ and substituting for ΔP gives

$$\frac{dy}{dt} = -\frac{\gamma_{lv}\, \cos(\theta - \alpha)\, x_0}{3\eta} \left(\frac{1}{y} - \frac{1}{y_0} \right),$$

which can be integrated to give

$$\frac{\gamma_{lv}\, \cos(\theta - \alpha)}{3\eta}\, \frac{x_0}{y_0}\, t = y_0\, \log\!\left(\frac{y_0}{y_0 - y} \right) - y. \tag{3.8}$$

Equation (3.8) may be used to calculate the order of magnitude of the time that it would take a moderately viscous adhesive to fill the crevices

Fig. 3.1. Penetration of an adhesive into a V-shaped groove.

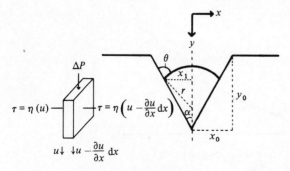

which might be found on a normal industrially prepared surface. It is obvious that equation (3.8) suggests that the crevice can never be filled completely, but the time to fill it to a depth of 99% of y_0 can be estimated by putting reasonable values into equation (3.8), such as $\gamma_{lv} \cos (\theta - \alpha) = 20$ mN m^{-1}; $\eta = 10^3$ N s m^{-2}; $y_0 = 10^{-4}$ m; $x_0 = 5 \times 10^{-3}$ m; then $t = 36$ s. This, of course, is rather longer than can be allowed in many industrial applications.

The above calculation is grossly oversimplified in that it neglects the back pressure of air trapped in the crevice and also the fact that pressure may be applied to assist the spreading of the adhesive. It is, however, of interest to calculate the relative effects of applied pressure and of surface forces in assisting the spreading of an adhesive. In the above example it can be seen that the effective pressure (given by equation (3.7)) when the crevice is half full is 800 Pa, so that quite a small applied pressure can accelerate the wetting of the surface and hence improve the joint strength.

From table 1.1 it can be seen that a typical value for the temperature coefficient of the surface tension $[(1/\gamma)\,(\partial\gamma/\partial T)]$ is $-0.002\ ^{\circ}$C^{-1}. A typical value for the temperature coefficient of the viscosity is $-0.04\ ^{\circ}$C^{-1}. The increase in joint strength with temperature of joint formation, shown in figure 3.2 for the system polyethylene-steel, is, therefore, more a function of the decrease in the viscosity of the adhesive than of the decrease in the surface tension.

Although flaws at an interface are undoubtedly a cause of joint weakness, they are not the only cause, and in many cases not even the major cause. Since, however, the actual effect of the flaws can only be considered

Fig. 3.2. Variation of joint strength for a stainless-steel–polyethylene lap joint with temperature of joint formation.

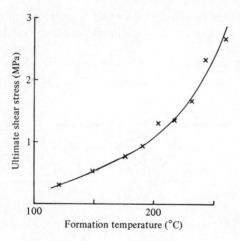

after the fracture mechanics of adhesive joints, further discussion will be deferred until a later chapter.

3.4 Adhesives

In section 3.1 it was concluded that 'an adhesive is any liquid which is capable of wetting a solid and subsequently solidifying in such a way that the shrinkage stresses induced by the change of state do not disrupt the joint'. However, for any given practical application an adhesive may have to meet a large number of specific requirements with regard to surface properties, viscosity, temperature stability, etc. A commercial adhesive is likely, therefore, to contain in addition to the base polymer or co-polymer a large number of additives which may modify some of the above properties to make the adhesive suitable for its specific application. It is beyond the scope of a monograph of this nature to give details of the various types of glue available; large volumes (Cagle, 1973; Shields, 1976; Patrick, 1969) have been written on the subject. However, in what follows the major classes of adhesives will be identified in order to demonstrate how the principles established in earlier chapters are utilised in practice.

A convenient classification for adhesives is on the basis of the mechanism by which the adhesives solidify, whether by evaporation of solvent, by chemical reaction, or by cooling through the melting point. Since this classification might exclude those adhesives which do not undergo a distinct phase change, but either become more and more viscous as the temperature drops, or simply owe their utilisation as an adhesive to the fact that at the operational temperature, they are sufficiently fluid to be able to form a coherent interface with a substrate, but are also sufficiently viscous to resist separation (the pressure-sensitive adhesives), these will all be considered in the last category.

Evaporation of solvent

Adhesives which solidify by evaporation of solvent may be further subdivided into those for which the solvent (or dispersion medium in the case of an emulsion) is water, and those for which it is an organic solvent.

The water-based adhesives are by far the oldest known class of adhesives. They consist of solutions or suspensions of high molecular weight compounds, usually of natural origin. Animal glues are normally prepared from skin, bone or sinew, and consist of solutions of the protein 'collagen' derived from the protein of animal blood or fish skins. Casein glue is based on a phosphoprotein derived from milk, and soya bean glue is based on the protein which can be abstracted very cheaply from soya beans.

High molecular weight compounds of vegetable origin similarly find very great application in the formation of water-based glues. Starches, dextrins, and related products occupy a very considerable portion of the market for adhesives, and in this class must also be considered latices based on both natural and synthetic rubber, which each form the basis of adhesive systems.

Finally, synthetic high-polymer emulsions and solutions can be prepared from polyvinyl alcohol, polyvinyl acetate or polyvinyl chloride, and mixtures or co-polymers of the basic polymers.

Because water is the basis of these glues, they are used primarily in situations where the porous nature of the substrate allows easy removal of the water, and on substrates which do not have such a low critical surface energy that wetting is inhibited. The addition of surfactants can lower the surface tension of the latex to about 40 mN m^{-1}, but for substrates with a lower critical surface tension other adhesive systems are to be preferred. Thus, the primary uses for the animal glues are as adhesives for wood, in applications such as furniture or musical instruments, leather, wood and paper in book-binding operations, and as binders for abrasive materials to a supporting cloth. Casein glues are used for wood and for paper coatings and some sticky labels.

Starches, dextrins and some of the synthetic glues find particular application in the glueing of paper, paper bags, envelope gums, wallpaper pastes, book-binding, and the formation of corrugated paper or paper tubes. A great deal of rubber latex, both natural and synthetic, is used in carpet-backing applications.

The second class of adhesives which solidify by loss of solvent are those based on volatile organic solvents. By far the most important of these are the adhesives based on various rubbers. Because such systems are based on non-polar hydrocarbon solvents, the surface tension of the adhesive is lower than that of water-based adhesives and good bonds can be formed with less polar surfaces. Solutions of natural rubber are widely used in such applications as tile cements (ceramic, rubber or PVC tiles), since in the absence of oxygen and sunlight chemical degradation of the rubber is unlikely. A solution of polychloroprene is used as a general purpose adhesive for fabrics and plastics, and is particularly widely used in footwear for joining synthetic soles to leather uppers. The more polar solvent-based rubbery adhesives are in general stronger, but have poorer wetting qualities. Acrylonitrile-butadiene rubbers are often used as adhesives for joining plastics to metals or in metal-to-metal applications.

Curing by chemical reaction

The rubbers described in the previous section were not vulcanised, or at least were not vulcanised subsequent to the formation of the joint. If some form of cross-linking can be introduced into the system subsequent to the formation of the joint then, since rupture must now involve the scission of covalent bonds, the strength of the joint will be greatly enhanced. Those systems which have been evolved involve either mixing the vulcanising agent with the rubber just before use or, alternatively, making a two-component system in which one component consists of the rubber formulation without the vulcanising agent and is applied to one surface, and the other component consists of the rubber formulation without the accelerator and is applied to the other surface. When the two surfaces are joined, cure can take place. (Wake, 1965, p. 394).

Most of the traditional thermosetting resins find application as adhesives. Any polymer which is highly cross-linked will have a high modulus and high strength, but will in general have a low elongation to break and hence a poor 'peel' or tear strength. Some of the thermosetting resins are therefore modified by the incorporation of elastomeric materials to introduce some flexibility without markedly reducing the strength.

Phenolic resins based on the condensation of phenol and formaldehyde are used widely in the preparation of plywood, and resorcinol/phenol formaldehyde resins, which can be cold cured and are therefore more suited to applications outside a factory, are similarly used for glueing wooden structures where high strength and resistance to high temperatures is required. When modified with nitrile rubber, the phenol formaldehyde rubbers have sufficient flexibility to be used in such applications as the bonding of metals and the bonding of brake linings in cars. A phenol formaldehyde resin modified with polyvinyl formal has been manufactured under the trade name 'Redux', and has been widely used as an adhesive for joining metals in aircraft structures; it is now being used as the adhesive in honeycomb sandwich structures in a number of aerospace applications.

Urea formaldehyde resins are widely used in the manufacture of chipboard, but the water sensitivity of the system is greatly reduced if the urea is at least partially replaced by melamine.

An important factor controlling the strength of a joint is the shrinkage stresses introduced by the curing of the adhesive. The resins based on the phenol or amino formaldehyde condensations all cure with the elimination of a molecule of water. Systems which cure without splitting out a product would be expected to have a smaller shrinkage on curing, and to form stronger bonds.

Epoxide resins based on the reactions of an epoxy group containing polymer (usually the diglycidyl ether of bis phenol A or one of its higher molecular weight homologues) with amines, amides, acids or phenols, have a very low shrinkage (<2% by volume) and consequently are very good structural adhesives. Because the resins are polar and are capable of forming polar interactions with a polar substrate, epoxy adhesives are widely used for joining metals, ceramics and glasses.

An alternative method by which a resin can cure without the elimination of a product is by the addition of a catalyst to a monomer capable of undergoing polymerisation. Unsaturated polyesters have been used for bonding plastics, but a more spectacular example of the addition poly-merisation of a monomer is provided by the 2-cyano-acrylates. Methyl-2-cyano-acrylate polymerises spontaneously when spread thinly on the surface of the adherends, due to the catalytic action of water or other weak bases adsorbed on their surface. These give very strong bonds, and higher homologues of the 2-cyano-acrylates are now being used as surgical adhesives to join tissue or bone.

Curing by phase transition

Thermoplastic adhesives, which solidify on cooling, currently probably account for the smallest quantity of adhesive used of the three major groups, but have the greatest capacity for development in the light of their ease of applicability in high-speed operations. Because they lack cross-linking, such adhesives tend to be weaker and more susceptible to creep, and so such adhesives are commonly used in less rigorous appli-cations, such as packaging. The common term for such adhesives is 'hot-melt adhesives'.

Polyvinyl acetals find use as hot-melt adhesives, polyvinyl formal being much used for the bonding of metals and polyvinyl butyral as the lami-nating agent for safety glass.

The polyamides, which are hard solids at ambient temperatures, have relatively sharp melting points which make them particularly easy to use in mass production machinery, and so they have been used for bonding such substrates as polyester film, aluminium foil and polyvinylidene chloride sheet. Polyethylene is, of course, a polymer which can easily be melted and then cooled through the melting point, and a particularly interesting application has been the formulation of an adhesive consisting of a mixture of iron powder and polyethylene which can be extruded as a thin strip, placed between non-conducting surfaces and then melted by induction heating to bond the surfaces together.

The so-called 'pressure-sensitive adhesives' will be considered in this class; these are usually adhesives which can wet the adherend and have a sufficiently low viscosity for small displacements so that they can conform to the irregularities of the substrates in a short time under a low stress. These adhesives are used for tapes which can be applied and stripped many times, and usually consist of unvulcanised elastomers to which have been added 'tackifiers', resins which reduce the viscosity and the relaxation time for the adhesive. The detailed analysis of the stripping of such a pressure-sensitive tape has been carried out by Kaelble (1971).

4.1   The creation of surfaces by fracture
Irreversibility of fracture

Whereas chapter 2 was concerned with the formation of surfaces and interfaces by such processes as wetting or by nucleation and growth, this chapter is concerned with their formation by fracture, either of an interface (adhesive fracture) or of a bulk solid (cohesive fracture). This chapter will be concerned, therefore, with such processes as the cracking of a piece of polymethyl methacrylate or the peeling of a piece of sticky tape from a solid substrate. It is in order to emphasise the basic similarity of such superficially different operations that the words cleavage, fracture or cracking, commonly associated with brittle solids, will in this chapter be used to refer to all processes in which new surfaces or interfaces are formed by the separation of previously adjacent molecules, irrespective of the nature of the material. Cleavage may thus refer to processes referred to in other texts as tearing, rupture etc.

The essential difference between the processes considered in chapter 2 and the processes involved in cleavage is that whereas the former may be considered as thermodynamically reversible processes, the latter are highly irreversible. Processes which involve the irreversible dissipation of considerable amounts of energy nearly always accompany the formation of new surfaces or interfaces by cleavage, and it is the objective of this chapter to consider the mechanisms involved in this irreversible dissipation of energy and the magnitude of the amounts involved.

Because of the existence of the energy dissipative processes which accompany cleavage, it is necessary to differentiate between the work of adhesion, W_a, defined by equation (2.1a) as

$$W_a = \gamma_1 + \gamma_2 - \gamma_{12}, \tag{2.1a}$$

which represents the *theoretical* minimum reversible work which would be necessary to separate two surfaces in the absence of these dissipative processes, and R_a, the adhesive fracture surface energy which represents the *actual* work necessary and includes all that which is dissipated irreversibly. Similarly the work of cohesion, W_c, defined by equation (2.1b) as

$$W_c = 2\gamma_1, \tag{2.1b}$$

must be distinguished from the cohesive fracture surface energy, R_c.

Adhesive and cohesive failure

It was because of a failure to recognise the difference between the reversible energy changes involved in interface formation and the irreversible energy changes involved in interface disruption, that for some years it was thought that a thermodynamic argument existed for the impossibility of true adhesive fracture. The argument ran as follows: if the adhesive (a) wets the substrate (s), then

$$\gamma_{sv} > \gamma_{av} + \gamma_{as}.$$

Hence, the work necessary to bring about adhesive fracture W_a, given by

$$W_a = \gamma_{av} + \gamma_{sv} - \gamma_{as},$$

must be greater than the work necessary to bring about cohesive fracture W_c, given by

$$W_c = 2\gamma_{av},$$

and hence a joint will always fail cohesively rather than adhesively. The argument is of course fallacious for all systems of practical interest since it is the fracture surface energy which controls joint breaking, whereas it is the thermodynamically reversible surface energy which controls joint making. Whilst there may be good grounds for doubting the probability of true adhesive failure (Bikerman, 1968, p. 137), the thermodynamic argument is not one of them.

4.2   The theory of fracture processes
The role of pre-existing cracks

The formation of new surfaces by cleavage involves both the initiation and the propagation of a crack through the material, and the energy requirements for these two processes may be very different. In this chapter the problem of crack initiation will not be discussed further, and the reader interested in this facet of fracture studies is referred to the review on this topic published by Kambour (1973).

The propagation of a pre-existing crack can only occur if two conditions are simultaneously fulfilled. There must be sufficient energy available to create the new fracture surfaces, and the stresses at the crack tip must be sufficiently large to overcome the intermolecular forces and disrupt the material. Concentration by different authors on one or other of these conditions for fracture has led to two apparently dissimilar approaches to the study of fracture. In what follows the two approaches will be termed the energy balance approach and the fracture mechanics approach, respectively, and it will be seen that under some circumstances the two approaches are precisely equivalent. There are, however, circumstances when the two approaches predict different values for (say) the applied force which will bring about fracture, and it is the delineation of these special circumstances which is of interest in determining the engineering utilisation of many plastics.

The energy balance approach

The energy required to create fracture surfaces is dissipated by a number of mechanisms. Some energy (usually a very small fraction of the total) appears as surface energy, as discussed in chapter 1. Usually a major portion of the energy input to the crack propagation process is utilised in plastic deformation of the material at the crack tip, which results in a very thin 'skin' of deformed material (Berry, 1972). Other mechanisms such as the emission of sound waves (Speake & Curtis, 1976) or even electromagnetic radiation (Deryaguin & Smilga, 1970) are possible. The term 'fracture surface energy' is used to denote the total energy utilised (either reversibly as surface energy or dissipated irreversibly) in the formation of unit area of crack, irrespective of the nature of the surfaces. It should be noted that unit area of a crack usually involves two surfaces, and the fracture surface energy, R, includes the energy required for both of them. In this chapter 'R' will be used to describe what in other texts may be termed surface work, tearing energy, adhesive failure energy, fracture toughness etc., and which may be given the symbols T, T, θ, etc., according to the circumstances of its determination.

The value of the fracture surface energy will vary according to the processes occurring at the crack tip. In this chapter attention will be concentrated on cracks which propagate in the so-called 'crack opening mode' (otherwise known as mode I), i.e. the surfaces of the crack move apart in directions perpendicular to the plane of the crack. Readers are referred to more specialised texts (e.g. Knott, 1973, particularly pp. 55-61) for a treatment of other modes of crack propagation.

The strain energy release rate

Since a certain amount of energy is required for the creation of an area of fracture surface, then a crack can only propagate if this amount of energy, or more, is made available for crack propagation (Griffith, 1920). This energy may derive from the action of the applied forces or, if the system is held at a constant strain, it may derive from the relief of the elastic stresses in the body which occurs as the crack spreads.

The energy which becomes available for crack propagation, either from the action of applied forces or by the relief of elastic stresses, may be calculated by a method derived by Gurney & Hunt (1967). If any system which contains a crack is acted upon by a force X, the points of application of which have been displaced by a distance u (figure 4.1), then extension of the crack will lead in general to a change in the stored elastic strain energy U of the system and an amount of work dw being done by the external force. The difference between the work done on the system by the external forces and the increase in the stored elastic energy of the system must be the energy available for the formation of crack surfaces. Irwin (1957) termed the amount of energy made available by extension of the crack by unit area the crack extension force, and gave it the symbol $\mathscr{G}$. Hence,

$$\mathscr{G}\,dA = dw - dU. \tag{4.1}$$

Now $dw = X\,du$, and for a linearly elastic system $U = \frac{1}{2}Xu$, hence substituting in equation (4.1)

$$\mathscr{G}\,dA = X\,du - \tfrac{1}{2}(X\,du + u\,dX)$$
$$= \tfrac{1}{2}(X\,du - u\,dX). \tag{4.2}$$

Now the compliance of a linearly elastic system is defined by $C = u/X$, i.e.

$$\frac{dC}{dA} = \frac{X\,du/dA - u\,dX/dA}{X^2},$$

whence

$$\mathscr{G} = \frac{1}{2}X^2\frac{dC}{dA}. \tag{4.3}$$

An alternative form of equation (4.3) may be derived by noting that

Fig. 4.1. Force acting on a cracked structure.

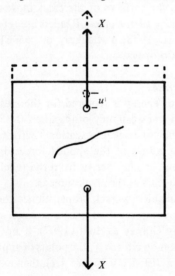

$$\frac{\mathrm{d}(1/C)}{\mathrm{d}A} = \frac{u\,\mathrm{d}X/\mathrm{d}A - X\,\mathrm{d}u/\mathrm{d}A}{u^2},$$

which may be combined with equation (4.2) to yield

$$\mathscr{G} = -\frac{1}{2}u^2 \left[\frac{\mathrm{d}(1/C)}{\mathrm{d}A}\right]. \tag{4.4}$$

Also, since X is a function of both A and u,

$$\mathrm{d}X = \left(\frac{\partial X}{\partial A}\right)_u \mathrm{d}A + \left(\frac{\partial X}{\partial u}\right)_A \mathrm{d}u,$$

and since for a linear elastic system $(\partial X/\partial u)_A = X/u$,

$$u\left(\frac{\partial X}{\partial A}\right)_u = u\frac{\mathrm{d}X}{\mathrm{d}A} - X\frac{\mathrm{d}u}{\mathrm{d}A} = u^2\frac{\mathrm{d}(1/C)}{\mathrm{d}A}.$$

Now

$$u\left(\frac{\partial X}{\partial A}\right)_u = 2\frac{\partial}{\partial A}\left(\frac{1}{2}uX\right)_u = 2\left(\frac{\partial U}{\partial A}\right)_u,$$

hence from equation (4.4)

$$\mathscr{G} = -\left(\frac{\partial U}{\partial A}\right)_u. \tag{4.5}$$

Equations (4.3), (4.4), and (4.5) are alternative definitions of the crack extension force, and from equation (4.5) can be seen the source of the alternative name for this quantity, the 'fixed grip strain energy release rate'.

The fracture criterion

The crack will only extend when $\mathscr{G}$ is equal to or exceeds the fracture surface energy, R. This value of $\mathscr{G}$ which is equal to R is given the symbol $\mathscr{G}_c$, and called 'the critical strain energy release rate'. The criterion for fracture in any system must then be that

$$\mathscr{G} \geqslant \mathscr{G}_c = R. \tag{4.6}$$

Consequently, if X_c is the minimum force which will just cause a crack to propagate, combining equations (4.6) and (4.3) gives the relationship

$$R = \frac{1}{2}X_c^2\frac{\mathrm{d}C}{\mathrm{d}A}, \tag{4.7}$$

which indicates that the force necessary to cause crack propagation is not only a function of the fracture surface energy of the material, but also the geometry of the system and the modulus of elasticity of its various parts, since these will affect $\mathrm{d}C/\mathrm{d}A$.

The fracture mechanics approach

Before a crack can propagate, the stress at the crack tip must be sufficient to break the intermolecular forces which hold the material together. In order to calculate the crack tip stresses, Irwin (1958) used a method of analysis devised by Westergaard (1939) to deduce the following equations for the stresses at the tip (T_1) of the crack shown in figure 4.2:

$$\left.\begin{array}{l} \sigma_y = \dfrac{K_I}{(2\pi r)^{1/2}} \cos(\theta/2) \left[1 + \sin(\theta/2)\sin(3\theta/2)\right] \\[2ex] \sigma_x = \dfrac{K_I}{(2\pi r)^{1/2}} \cos(\theta/2) \left[1 + \sin(\theta/2)\sin(3\theta/2)\right] \\[2ex] \sigma_{xy} = \dfrac{K_I}{(2\pi r)^{1/2}} \sin(\theta/2) \cos(\theta/2) \cos(3\theta/2). \end{array}\right\} \quad (4.8)$$

In these equations σ_x, σ_y and σ_{xy} are the normal and shear stresses in the directions shown in figure 4.2 at the point P whose polar coordinates, based on the crack tip as origin, are r and θ. The term K_I depends on the applied stress and on the geometry of both the crack and the rest of the system; it is known as the stress intensity factor, and expressions for K_I for several standard geometries have been tabulated (e.g. Knott, 1973; Paris & Sih, 1965). The subscript 'I' after the K refers to mode I crack propagation, which is the only mode for which equations (4.8) are valid. A number of interesting features may be noted about equations (4.8). The form of the stress field is independent of the applied stress, only the magnitude of the stress field is affected by changes in the applied stress. The crack tip stresses rise as the crack tip is approached: as r approaches zero, σ appears to approach infinity, but of course this does not happen because in the region where the stresses would, according to equations (4.8), have exceeded the yield stress of the polymer, flow takes place and σ assumes a more or less constant value over the whole

Fig. 4.2. Calculation of strain energy release rate.

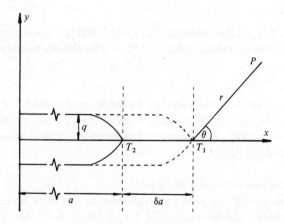

of this region. In general, the 'plastic zone' where yield has taken place is small enough to be neglected in many calculations. Only the first term in a series expansion has been included in equations (4.8), which means that the expressions are only valid when r is much smaller than the length of the crack.

The failure criterion

Crack propagation is assumed to occur when the stress at the crack tip reaches a critical value at which molecular bonds are broken. The crack tip is, however, in the 'plastic zone' where equations (4.8) do not apply; but since it may be assumed that there is a unique relationship between stresses in the plastic zone and stresses in the elastic region for which equations (4.8) do apply, it can be seen that the assumption of a critical stress at the crack tip is the same as the assumption of a critical value of the stress intensity factor in the elastic region. This critical value of the stress intensity factor is termed K_{Ic} (the I referring to mode I crack propagation), and the criterion for crack propagation is that

$$K \geqslant K_{Ic}.$$

The relation between the two criteria

Under certain circumstances the two different criteria for crack propagation can be shown to correspond to the same stress condition in the body. Irwin (1957) suggested that if the crack propagation process is reversible in the sense that no energy is lost to the system when the crack propagates, then under fixed grip conditions when no work is done by externally applied forces, the energy which is made available for the creation of fracture surfaces by the propagation of a crack a distance δa must be equal and opposite to that which must be supplied to the system to close the extended crack the same distance δa. The work necessary to close a crack is calculated quite simply. When the crack is closed to the position shown by the full line in figure 4.2, then the stress σ^* on the x-axis acting in the y-direction is given by equation (4.8) with $\theta = 0$ and r measured from T_2. When the crack is open, in the position shown by the broken line in figure 4.2, the stresses in the y-direction on the crack surfaces must be zero. Consequently, if the system behaves in a Hookean manner, the work done in bringing together a section of crack of length dx and unit width is $\frac{1}{2}\sigma^* q dx$, where q is the original distance of separation of the crack surfaces as shown in figure 4.2.

Now Westergaard's treatment of the stresses and strains at the tip of a stressed crack also gave rise to an expression for the shape of the crack surfaces as they approach the tip. Using the notation of figure 4.2 this gives for the crack tip displacement in plane strain

$$q = \frac{K}{(\pi a)^{1/2}} \frac{2(1 - \nu^2)}{E} (a^2 - x^2)^{1/2}, \tag{4.9}$$

where E is the Young's modulus and ν is Poisson's ratio for the material. Hence, the work done in closing a crack a distance δa is given by

$$dw = 2 \int_0^{\delta a} \tfrac{1}{2} \sigma^* q \, dx,$$

where

$$\sigma^* = \frac{K}{(2\pi r)^{1/2}}$$

from equation (4.8), and

$$q = \frac{2K(1 - \nu^2)}{(\pi a)^{1/2} E} \left[(a + \delta a)^2 - (a + r)^2 \right]^{1/2}$$

$$= \frac{2K(1 - \nu^2)}{(\pi a)^{1/2} E} \left[2a(\delta a - r) \right]^{1/2},$$

neglecting small terms. Hence, since along the x-axis $dx = dr$,

$$dw = \frac{2(1 - \nu^2)}{\pi} \frac{K^2}{E} \int_0^{\delta a} \left(\frac{\delta a - r}{r} \right)^{1/2} dr.$$

This can be integrated to yield

$$dw = \frac{K^2 (1 - \nu^2)}{E} \delta a,$$

and since, by the definition of fixed grip strain energy release rate $\mathcal{G}$, $dw = \mathcal{G} \delta a$,

$$\mathcal{G} = \frac{K^2 (1 - \nu^2)}{E}. \tag{4.10}$$

Equation (4.10) implies that provided the conditions at the crack tip can be described by equations (4.8) and (4.9), then the critical stress intensity factor corresponds to a critical strain energy release rate, and the two crack propagation criteria are one and the same thing.

4.3   Experimental determination of the fracture surface energy

Methods for the determination of the fracture surface energy may be arbitrarily divided into those which have been used with soft elastomeric materials and those which have been used with hard brittle materials. In this chapter greater attention will be paid to methods for the determination of the fracture surface energy of an interface than of a homogeneous solid, since good treatments of the fracture of polymeric solids have been prepared recently by Berry (1972), Andrews (1968), and Gent (1972).

Methods for brittle materials

Although a number of methods have been developed for determining the fracture surface energy of brittle solids or non-flexible adhesive joints, attention will be concentrated here on those systems which involve quasi-static crack propagation. The great advantage of such systems is that, because the crack is controllable, a large number of determinations of the fracture surface energy are possible from each specimen.

The double cantilever beam arrangement shown in figure 4.3 has been utilised by a number of workers in this field; Berry (1963) used specimens in which the crack was constrained to move along the axis by means of a groove which was machined there. Gurney & Hunt (1967) have shown on thermodynamic grounds that in the absence of such a groove the crack will normally run off to one side of the specimen.

If each arm of the double cantilever beam is considered as a simple built-in beam with a force X applied to its end, then the deflection of the end is given by

$$\frac{u}{2} = \frac{Xa^3}{3EI} = \frac{4Xa^3}{Ebh^3}, \qquad (4.11)$$

where E is Young's modulus for the material, I is the moment of inertia, and the significance of the other symbols is shown in figure 4.3. The compliance of the specimen, $C = u/X$, is therefore given by

$$C = 8a^3/Ebh^3.$$

From equation (4.7)

$$R = \frac{1}{2}X_c^2 \frac{dC}{dA}.$$

If the thickness of the section along which the crack propagates is t,

Fig. 4.3. Double cantilever beam arrangement for determination of fracture surface energy.

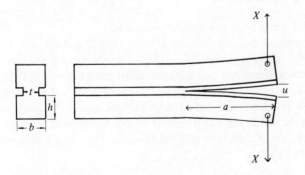

$$\frac{dC}{dA} = \frac{dC}{da}\frac{da}{dA} = \frac{1}{t}\frac{dC}{da},$$

Therefore,

$$R = \frac{12X_c^2 a^2}{Ebth^3},$$

and so the fracture surface energy can be determined from a knowledge of the force required to cause a crack of known length to propagate.

The simple beam formula represented by equation (4.11) is not, however, an adequate description of the behaviour of an arm of a double cantilever beam, for a number of reasons. It neglects the deflection of the beam due to shearing forces, the effective length of the beams is longer than 'a' due to rotation of the ends which were assumed to be 'built-in', and the modulus E is of course time dependent for viscoelastic polymeric materials. Berry (1963), therefore, adopted a slightly different treatment; he assumed that the elastic behaviour of the system could be described by an expression of the form

$$X = \phi a^{-\lambda} u, \tag{4.12}$$

where ϕ and λ are arbitrary constants, which may be determined from subsidiary experiments in which the slope of the force-deflection curve ($m = X/u$) is determined as a function of crack length. These subsidiary experiments can be carried out at small strains, well below those at which significant crack propagation can occur, and from the relationship

$$\log m = \log \phi - \lambda \log a,$$

both ϕ and λ are easily obtained. The combination of equations (4.12) and (4.7) then yields

$$R = \frac{1}{2}\frac{X_p^2 \lambda a^{\lambda-1}}{\phi t}$$

if the force necessary to cause crack propagation (X_p) is determined, or

$$R = \frac{1}{2}\frac{X_p u\lambda}{ta}$$

if both force and beam deflection are measured.

The method used by Gurney & Hunt (1967) and Gurney & Amling (1970) for the determination of the fracture surface energy of both homogeneous specimens and adhesive joints is even simpler. Their experimental arrangement and a typical force-deflection curve are shown in figure 4.4. If ΔA represents the change in crack area as the specimen moves from the state represented by point A to that represented by point B, then dw = area $ABB'A'$ and dU = area OBB' − area OAA', hence from equations (4.1) and (4.6)

$$R\Delta A = ABB'A' - (OBB' - OAA')$$
$$= OAB,$$

and the fracture surface energy may be simply determined from the area of the triangle OAB and the crack lengths at A and B.

In order to avoid the difficulties associated with the necessity of monitoring the load and the length of the crack simultaneously, Ripling, Mostovoy & Corten (1971) introduced a specimen which, because the strain energy release rate, $\mathscr{G}$, was for any given load independent of crack length, would require measurements of the crack propagation force alone to determine the fracture surface energy.

Equation (4.11) describes the displacement of the ends of the cantilever beam due to the action of the tensile forces acting in the chords of the beam. The force X also applies a shear stress to the beam so that if the angular deflection is θ then

$$\theta = \frac{X/bh}{G}$$

where G is the shear modulus. If $\frac{1}{2}u'$ is the additional deflection due to the action of shear forces $\frac{1}{2}u'/a = \theta$, i.e.

$$\frac{u'}{2} = \frac{aX}{bhG}. \tag{4.13}$$

In general, for many brittle materials G may be put equal to $\frac{1}{2}E$, and so adding u' of equation (4.13) to u of equation (4.11) gives for the compliance of the specimen

$$C = \frac{8a^3}{Ebh^3} + \frac{4a}{Ebh}.$$

Equation (4.3) gives

$$\mathscr{G} = \frac{1}{2}X^2\frac{\mathrm{d}C}{\mathrm{d}A}$$

Fig. 4.4. Determination of the fracture surface energy of an adhesive joint.

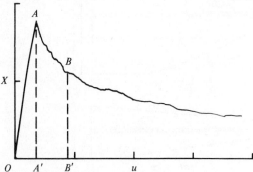

Therefore,

$$\mathscr{G} = \frac{4X^2}{Ebt} \left(\frac{3a^2}{h^3} + \frac{1}{2h} \right).$$

Now, Ripling *et al.* have suggested that if the specimen is contoured so that

$$\left(\frac{3a^2}{h^3} + \frac{1}{h} \right) = \text{const} = m',$$

then the strain energy release rate is independent of the length of the crack. Such a specimen is termed a tapered double cantilever beam specimen. Examples of such specimens are shown in figure 4.5 for a homogeneous specimen and an adhesive joint.

An alternative technique for the study of crack propagation under constant strain energy release rate conditions is by use of the 'double torsion' test, shown in figure 4.6 (Evans, 1972). Rotation of the 'arms' on either side of the cracked region under the action of the applied stress causes the crack to propagate along the line of the machined groove in the bottom of the specimen. Although the crack propagates along the axis of the specimen, the actual motion of the crack surfaces at the crack tip is perpendicular to the plane of the crack, and so in the double torsion test mode I crack propagation is being studied. An approximate analysis of the compliance of the specimen can be made by considering each arm of the specimen as a rectangular shaft (Timoshenko, 1955, p. 289). Using the nomenclature of figure 4.6, the angle of twist per unit length of each arm, θ, is given by

Fig. 4.5. Tapered double cantilever beam arrangement for the determination of the fracture surface energy of (*a*) Bulk specimen, (*b*) Joint specimen.

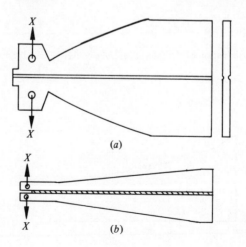

$$\theta = \frac{2M_t}{\beta bc^3 G},$$

where M_t is the torque and equals $\frac{1}{2}Xd$, since the load is shared equally between the two arms; β is a constant which approaches $\frac{1}{3}$ for wide specimens, and G is the shear modulus of the material. Hence, if y is the deflection of the inside edge of the arm and a is the crack length, $y = a\theta d$.

Now if D represents the compliance of the uncracked material and if C is the total compliance,

$$C = \frac{y}{X} + D = \frac{3ad^2}{bc^3 G} + D,$$

i.e.

$$C = Ba + D, \tag{4.14}$$

where $B = 3d^2/bc^3 G$. Since dC/da is a constant, cracks propagate under a constant strain energy release rate, and therefore at constant velocity, V.

From (4.14), the deflection y is given by

$$y = X(Ba + D);$$

consequently, differentiating with respect to time gives

$$S = \frac{dy}{dt} = (Ba + D)\frac{dX}{dt} + XB\frac{da}{dt}$$

$$= (Ba + D)\frac{dX}{dt} + BXV, \tag{4.15}$$

where S is the cross-head displacement rate of the testing machine, and $V = da/dt$ and is the crack propagation velocity. At constant crack velocity,

Fig. 4.6. The double torsion test for the determination of fracture surface energy. (From Evans, 1972, p. 1138.)

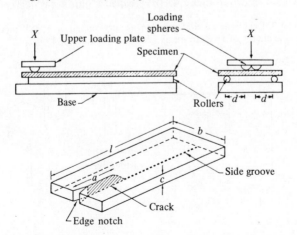

since the strain energy release is constant, X is a constant; consequently under constant cross-head speed conditions of testing, equation (4.15) reduces to

$$S = BXV, \tag{4.16}$$

which means that the double torsion test is particularly useful for the determination of the variation of R or K with crack velocity.

Methods for elastomeric materials

Rivlin & Thomas (1953) suggested the three experimental configurations shown in figure 4.7 for the determination of the fracture surface energy. They suggested that for a specimen of the form of figure 4.7(a) the presence of an edge cut would reduce the strain energy in the specimen. If this reduction is termed ΔU, then from the geometry of the area in which the strain energy is reduced,

$$\Delta U = K'a^2 W_b t,$$

where K' is either a constant or varies only slowly, W_b is the stored elastic energy of the bulk specimen and t is the thickness. If the specimen is subjected to tension in a stiff machine, so that cut extension takes place at constant displacement, then in equation (4.1) $dw = 0$ and $\mathcal{G} = R$, so

$$Rtda = dU = 2K'W_b tada,$$

or

$$R = 2K'W_b a.$$

For most elastomeric systems K' lies between 1 and 3 (Gent, 1972), and W_b can be determined from the work necessary to stretch a specimen which contains no cut to the same strain as that at which cut propagation was initiated.

The specimen shown in figure 4.7(b) is called a 'pure shear' test specimen

Fig. 4.7. Experimental configurations for determination of the fracture surface energy of an elastomer. (After Lake & Lindley, 1966, p. 177.)

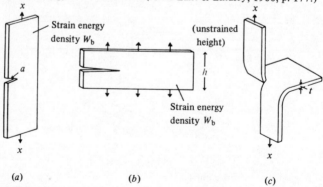

(a) (b) (c)

because the material in the region away from the influence of the cut is in a state of pure shear. In spite of this, the propagation of the crack causes the faces to move apart normal to the plane of the cut, and so this specimen also measures a mode I fracture surface energy. In the specimen shown in figure 4.7(b), extension of the cut by a distance da causes transfer of a volume of rubber $htda$ from the strained to the unstrained state. If this takes place at constant displacement, then substituting in (4.1), $dw = 0$, $\mathscr{G} = R$ and so

$$Rtda = W_b htda,$$

or

$$R = hW_b,$$

where W_b is the stored elastic energy per unit volume of the rubber at the strain at which the cut extends.

In the so-called 'trouser-leg' tear test shown in figure 4.7(c), stable cut propagation may take place at constant load, so that if e is the strain in the 'legs' of the specimen (usually deformations will be sufficiently small to allow the 'conventional strain' as a measure of deformation) then an increase in cut length of da will result in an additional separation of the ends by $2(1 + e)da$ and the transfer of a length da from the strained to the unstrained portion of the specimen. Hence, substituting in equation (4.1), remembering that during stable crack growth $\mathscr{G} = R$,

$$dw = 2X(1 + e)da,$$

$$dU = \frac{1}{2}\frac{2X}{ht}e\,ht\,da,$$

so that

$$Rtda = 2X(1 + e)da - Xeda,$$

or

$$R = \frac{2X}{t}\left(1 + \frac{e}{2}\right).$$

In general $e/2 \ll 1$ so that $R = 2X/t$.

The analogous methods for the determination of the fracture surface energy of an interface were developed from the methods of Rivlin & Thomas (1953) by Gent & Kinloch (1971) and Andrews & Kinloch (1973). These are shown in figure 4.8, and it is immediately evident that the fracture surface energy for the three systems is given by

$$R = \tfrac{1}{2}K''aW_b,$$

$$R = hW_b,$$

and

$$R = X/h,$$

for the simple extension, pure shear and peeling test pieces, respectively.

The test methods described for elastomeric systems have so far assumed stiff testing machines. However, with elastomeric systems crack propagation is often stable in a soft machine, and a good example of a soft machine is the rotating strip adhesiometer devised by Deryaguin (Moskvitin, 1969, p. 10), and illustrated in figure 4.9.

If a length da of the interface is separated, then for the terms in equation (4.1), since $\mathscr{G} = R$,

$$dw = X(1 + e)\, da\, (1 - \cos \alpha),$$
$$dU = \tfrac{1}{2} X\, e\, da,$$
$$Rb\, da = X(1 + e)\, da\, (1 - \cos \alpha) - \tfrac{1}{2} X e\, da,$$

where e is the extension in the detached adhesive. Usually $e \ll 1$ and so

$$R = \frac{X}{b}(1 - \cos \alpha). \tag{4.17}$$

The system is rotated until the adhesive starts to detach and then the fracture surface energy is determined from (4.17).

4.4 Stability of crack propagation

The stress intensity factor for a central crack of length $2a$ in a large flat plate stressed perpendicular to the plane of the crack is given by (Knott, 1973, p. 63)

$$K_I = \sigma(\pi a)^{1/2}, \tag{4.18}$$

Fig. 4.8. Experimental methods for determination of the fracture surface energy of joints involving flexible adhesives, (*a*) Simple extension test piece, (*b*) Pure shear test piece, (*c*) Peel test piece.

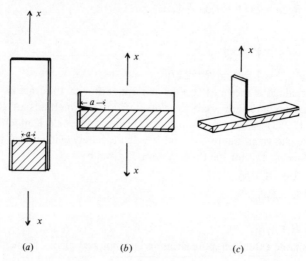

(*a*)   (*b*)   (*c*)

where σ is the applied stress. If equations (4.6), (4.10) and (4.18) are combined, this yields the result that the crack will propagate when

$$\sigma = \left[\frac{ER}{\pi(1 - \nu^2)a} \right]^{1/2}. \tag{4.19}$$

Equation (4.19) is a form of the 'Griffith equation' published originally by Griffith in 1920 (actually the original publication gave a slightly different expression, but this was corrected later). It indicates that under constant load conditions a crack which starts to propagate will (because as a increases, σ becomes less) accelerate and the whole system will fail in an unstable manner. This is a good example of an instability which is caused by the geometry of the testing system leading inevitably to the situation where $\mathscr{G} > R$. Under these circumstances the excess energy above that which is necessary for the formation of the surfaces may be transformed into kinetic energy of motion of the crack surfaces relative to one another. In this case the crack will accelerate and will continue to do so until it reaches a limiting velocity controlled only by the speed at which elastic energy can be transmitted through the material. This is termed unstable or rapid crack propagation. If, on the other hand, $\mathscr{G} = R$ then all the energy which is made available by the crack propagation process is utilised as fracture surface energy, and a negligible amount is transformed into kinetic

Fig. 4.9. Deryaguin's rotating strip adhesiometer.

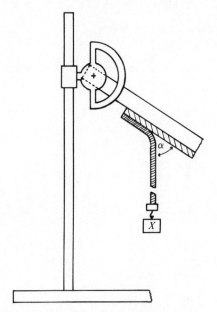

energy. The crack then propagates in a slow, stable or quasi-static fashion (Gurney & Hunt, 1967).

The tapered double cantilever beam and the double torsion test described previously were designed to ensure that crack propagation was stable, since the situation of an increasing strain energy release rate with increasing crack length could not arise with these experimental configurations. It was, however, found that for some polymers, such as polymethyl methacrylate in methanol (Hakeem & Phillips, 1978) or some epoxy resins in air (Yamini & Young, 1977) crack propagation is unstable and gives rise to a saw-tooth force–deflection curve as in figure 4.10, even in those tests for which the strain energy release rate does not increase with crack length.

There are basically three factors which may bias crack propagation towards instability. The first factor is a positive value for $d\mathscr{G}/dA$ which may arise from the geometry of the specimen. The second factor is a negative value for dR/dA which may arise from environmental effects, more particularly when a liquid can be adsorbed at the crack tip and locally increase the fracture surface energy; when the crack has traversed the toughened region it will reach the region of lower fracture toughness and propagation may become unstable (Mai, 1975). An alterna-

Fig. 4.10. Typical curves of load X against displacement y for epoxy resin double torsion specimens during crack propagation. (*a*) Continuous stable cracking, (*b*) Crack jumping or stick–slip behaviour. (After Yamini & Young, 1977, p. 1075.)

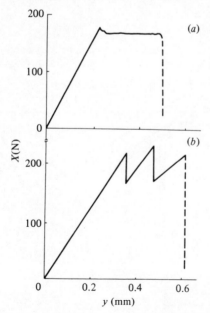

tive mechanism by which dR/dA may apparently become negative has been proposed by Cherry & Thomson (1978); they have suggested that if the specimen is deformed sufficiently slowly, then blunting of the crack tip will occur and this may reduce the crack tip stress to a much lower level than that which would be found at the tip of a sharp crack. Under these circumstances the strain energy release rate may well exceed the fracture surface energy before the crack tip stress is sufficient to rupture the inter-molecular bonds and initiate fracture. When the crack does start moving, then the excess strain energy released causes instability of crack propagation. The third factor which might bias crack propagation towards instability is a negative value of $dR/d\dot{a}$, where $\dot{a}$ is the crack velocity, which may arise when the fracture surface energy decreases with increasing crack velocity so that once a crack has started moving it accelerates in an unstable fashion. Marshall, Coutts & Williams (1974) have suggested that a negative value of $dR/d\dot{a}$ may be associated with adiabatic heating at the crack tip when the crack propagates at velocities which are too great to allow the dissipation of the heat by conduction before the crack propagates into the softened region.

The conditions for crack stability

The role of the various factors which play a part in determining the stability or otherwise of crack propagation will vary according to the type of fracture test which is being performed. In a so-called 'hard' or 'stiff' testing machine, the relative displacements of the points of application of the force always increase irrespective of the response of the system being tested, i.e. du/u is positive. In a soft testing machine the force always increases, irrespective of the changing relative displacements of the points of application of the force due to the response of the system being tested, i.e. dX/X is positive (Gurney & Hunt, 1967). The condition for crack stability is that $\mathscr{G} = R$; hence, substituting from equation (4.3)

$$R = \frac{1}{2} X^2 \frac{dC}{dA}.$$

Differentiating with respect to A gives

$$\frac{dR}{dA} - \frac{1}{2} X^2 \frac{d^2C}{dA^2} - X \frac{dX}{dA} \frac{dC}{dA} = 0,$$

but in a soft testing machine $dX/X > 0$ and, since dividing through by $X^2 dC/dA$ yields

$$\frac{dR}{dA} \Big/ X^2 \frac{dC}{dA} - \frac{1}{2} \frac{d^2C}{dA^2} \Big/ \frac{dC}{dA} = \frac{1}{X} \frac{dX}{dA}, \tag{4.20}$$

the left-hand side of (4.20) must be positive if crack propagation is to be stable. Hence, writing $2R$ for $X^2\,dC/dA$, from (4.7), the criterion for crack stability in a soft testing machine becomes

$$\frac{1}{R}\frac{dR}{dA} \geqslant \frac{d^2C}{dA^2}\Big/\frac{dC}{dA}. \tag{4.21}$$

In a similar fashion it can be shown (Gurney & Hunt, 1967) that the stability criterion for crack propagation in a stiff testing machine (du/u positive) can be derived by putting $\mathcal{G} = R$ in equation (4.4) and differentiating with respect to A as before. The result is that for stability

$$\frac{1}{R}\frac{dR}{dA} \geqslant \frac{d^2(1/C)}{dA^2}\Big/\frac{d(1/C)}{dA}, \tag{4.22}$$

which by simple algebraic manipulation yields

$$\frac{1}{R}\frac{dR}{dA} \geqslant \frac{d^2C}{dA^2}\Big/\frac{dC}{dA} - \frac{2}{C}\frac{dC}{dA}. \tag{4.23}$$

Since C and dC/dA must both be positive, a comparison of equations (4.22) and (4.23) shows that stability is easier to achieve in a stiff machine. Mai & Atkins (1975) following Gurney & Mai (1972) have termed the quantities on the right-hand sides of equations (4.21), (4.22) and (4.23) the geometric stability factors, since these quantities contain all the factors which are associated with the geometry and the elastic properties of the specimen. The factors which are concerned with the fracture properties of the system (R, dR/dA and dR/dA) are all contained in the term on the left-hand side of equations (4.21) and (4.23). From the form of equations (4.21) and (4.23) it can be seen that the geometric stability factors may be written as n_X/A for a load controlled machine and n_u/A for a displacement controlled machine. Obviously the smaller the algebraic value of n_X or n_u the greater is the chance of stable cracking.

The stability of various test pieces

The relative propensities of different test configurations for stability of crack propagation may be assessed from the values of n_u. If one considers the large flat plate stressed perpendicular to the plane of the crack, as in section 4.4, then from equations (4.18), (4.10) and (4.3), dC/dA can be shown to be $2\pi a(1 - \nu^2)/l^2 tE$ where l and t are the width and thickness of the plate and E is Young's modulus. Since for a large specimen C will be large, the last term in equation (4.23) is negligible and the geometric stability factor is $+1$.

From equation (4.12) it may be seen that for a simple double cantilever beam test configuration, $n_u = -(1 + \lambda)$, and since λ is often approximately 2.7, this implies that $n_u \approx -4$, so that the double cantilever beam test is inherently more likely to give stable cracking than the large flat plate with a centrally situated crack discussed above.

Although the tapered double cantilever beam test configuration offers the advantage of a constant value for the strain energy release rate throughout the range of crack lengths encountered, it is not a configuration which promotes stability. Mai (1974) has shown that for a tapered double cantilever beam specimen, n_u usually lies between -2 and -3 and so is slightly less prone to stability than the parallel sided specimen. However, if the specimen is forced to crack from the wide end towards the pointed end, then Mai & Atkins (1975) have shown that n_u decreases from a value of -4 for short cracks to values less than -18 for long cracks, so that cracking is inherently very stable.

Finally, it should be noted that the geometric stability factor for a double torsion configuration can be calculated from (4.14) to be $-2(1 + D/aB)^{-1}$, so that for long crack lengths the double torsion configuration is less prone to cause stable crack propagation than the double cantilever beam.

5   Fracture surface energy and strength

5.1   Work of cohesion and the cohesive fracture surface energy

Introduction

In chapter 4 methods were described by which the fracture surface energy for adhesive or cohesive fracture of both brittle and elastomeric systems could be determined. In this chapter the results obtained from the use of these techniques will be compared with the theoretical values which would be expected on the basis of the ideas put forward in chapters 1 and 2. The discrepancies which are found will form the basis of an enhanced understanding of the strength of both homogeneous solids and of adhesive joints.

The work of cohesion

It was shown in chapter 2 that the work of cohesion for a solid body is given by

$$W_c = \pi n^2 A / 16 r_0^2. \tag{2.2'}$$

This expression was based on the Lennard-Jones expression for inter-molecular energy

$$\epsilon = -A/r^6 + B/r^{12}. \tag{1.23}$$

The same expression can be used to calculate the magnitude of the reversible work of cohesion. From equation (1.23), if at the equilibrium inter-molecular spacing $(d\epsilon/dr)_{r=r_0} = 0$, the molecular interaction energy ϵ_0 is given by $\epsilon_0 = -A/2r_0^6$. Substitution in equation (2.2') yields

$$W_c = (\pi/8) n^2 \epsilon_0 r_0^4. \tag{5.1}$$

Now if it is assumed for polymethyl methacrylate that $n = 7 \times 10^{27}$ monomer units/m^3, that $\epsilon_0 = 3.6 \times 10^{-19}$ J/molecule and $r_0 = 1.6 \times 10^{-10}$ m, then $W_c = 36$ mJ m^{-2}, which is close to the value which might be determined by the surface chemical techniques of chapter 2.

Since the work of cohesion should represent the reversible work necessary for the creation of two fracture surfaces, it is of interest to compare the theoretical value of the cohesive fracture surface energy with actual values determined experimentally. Berry (1963), using a double cantilever beam technique, determined the fracture surface energy of a crack propagating in polymethyl methacrylate at temperatures between $-196\,°$C and $+50\,°$C, and found that it decreased from about 1.05 kJ m^{-2} at $-196\,°$C to about 0.19 kJ m^{-2} at $+50\,°$C. The discrepancy by a factor exceeding 10^4 between the theoretical reversible work of cohesion and the actual irreversible work of cohesion is attributed to the plastic work done at the crack tip as the crack propagates through the material leaving a thin skin of highly deformed material on each crack surface.

Similarly, Rivlin & Thomas (1953) determined the 'tearing energy', the fracture surface energy for a number of rubbers, by the techniques described in chapter 4. The values obtained were of the order of $10 \, kJ \, m^{-2}$. Unlike the brittle plastics examined by Berry, an elastomer should not be able to form a permanently deformed skin on the fracture surfaces since a perfect elastomer does not have a yield point. The discrepancy between the theoretical reversible work of cohesion in a rubber and the actual irreversible fracture surface energy is attributed to viscoelastic hysteresis losses taking place as the crack propagates.

For both plastics and elastomers, therefore, it can be seen that the experimentally determined fracture surface energies are vastly in excess of the theoretical reversible work of cohesion. However, since the processes which dissipate energy during crack propagation differ between two types of material, plastics and elastomers will be considered separately below.

Fracture surface energy of brittle plastics

The fracture surface energy is a function of the velocity at which the crack is propagating through the polymer. The variation of the fracture surface energy with velocity for polymethyl methacrylate has been plotted by Berry (1972) from data obtained by Vincent & Gotham (1966) using a variety of techniques. The results are shown in figure 5.1. The increase in fracture surface energy with increase in crack propagation rate, which can be seen at the lower rates in figure 5.1, can be ascribed to the increase in the energy which is necessary to bring about the deformation of the surface layers at the higher strain rates. The subsequent decrease in fracture surface energy at the higher strain rates has been attributed (Vincent & Gotham, 1966) to a change from isothermal to adiabatic conditions; at the higher crack propagation rates it is suggested that there is insufficient time for the heat generated by the plastic deformation of the surfaces to be conducted away, and the consequent heating of the material in the vicinity of the crack tip lowers the yield point and reduces the work of deformation.

Williams & Marshall (1975), using a model originally suggested by Dugdale (1960), developed a quantitative explanation of the form of figure 5.1. Dugdale's treatment was originally developed for a metal, and attempted to model the plastically deformed zone at the tip of the crack, as shown in figure 5.2 (Williams & Marshall, 1975). In the region where the stresses described by equations (4.8) would exceed the yield stress σ_c of the material, it is assumed to have yielded and this whole zone is supposed to be subjected to a uniform stress σ_c. Unlike metals, many polymers form crazes at the crack tip, but Brown & Ward (1973) have shown experimentally that for such polymers the craze shape is close to that which would be predicted by the Dugdale model if σ_c is the crazing stress. As a result of the crazing or yielding of the polymer, the crack tip is no longer infinitely sharp, but the crack surfaces are separated at the crack tip by a distance u_c. The normal term for u_c is the 'crack-opening displacement'. Hence, using the nomenclature of figure 5.2 it can be shown (Knott, 1973, p. 69) that

$$\mathscr{G}_c = u_c \sigma_c,$$

and

$$\Delta = \pi K_c^2 / 8\sigma_c^2,$$

from which, remembering equation (4.10),

$$\mathscr{G} = \frac{K^2(1 - \nu^2)}{E}, \tag{4.10}$$

it may be seen that

$$K_c^2 = u_c \sigma_c \, E/(1 - \nu^2). \tag{5.2}$$

Now, provided that the crack is propagating sufficiently quickly, the time taken for the crack to propagate through the Dugdale zone is given by $\tau = \Delta/\dot{a}$, where $\dot{a}$ is the crack propagation velocity. The time dependence of both the yield or craze strength of a polymer and its modulus may be written as a power law, so that in order to compare the effective values for

Fig. 5.1. Variation of fracture surface energy with crack velocity for poly-methyl methacrylate. (After Berry, 1972, p. 65.)

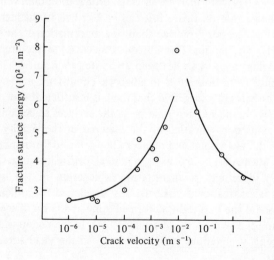

these terms at different rates of cracking it is possible to write

$$\sigma_c(\tau) = \sigma_0 \, \tau^{-m}$$

and   (5.3)

$$E(\tau) = E_0 \, \tau^{-n}$$

where m and n are constants and σ_0 and E_0 represent the effective yield/craze strength and modulus at a standard rate of cracking. Hence, if a failure criterion of a constant crack-opening displacement is assumed, the relationship between K_c and $\dot{a}$ may be written as

$$K_c = \text{const} \times (\dot{a})^{(m + n)/2(1 - m + n)}.$$

According to Williams & Marshall m and n may, for a wide range of polymers be approximated by 0.1, so that, approximately

$$K_c \approx \text{const} \, (\dot{a})^{0.1}.$$

It can thus be seen that the observed increase in fracture toughness, and hence in fracture surface energy, with increasing crack propagation velocity may be ascribed to the increase in modulus and yield/craze stress of the polymer with increasing strain rate.

Fig. 5.2. The Dugdale model of a crack tip. (From Williams & Marshall, 1975.)

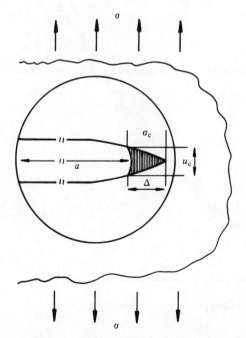

Fracture surface energy of elastomers

The effect of the rate at which a crack propagates through a series of rubbers is shown in figure 5.3 (Gent, 1972; Mullins, 1959). It should be noted that this curve was obtained by determining the fracture surface energy over a limited range of crack propagation rates at a number of temperatures for each of the rubbers examined, and then using a method similar to that developed by Williams, Landel & Ferry (WLF) (1955) in which the results are transposed by shifts along the log time axis to form a master curve at some reference temperature (in this case, 20 °C above the glass transition temperature). It can be seen from the coincidence of the curves for different rubbers that the tearing energy of these rubbers is affected by their chemical composition only as far as the latter affects the glass transition temperature.

The second feature that can be noted from figure 5.3 is that at normally accessible crack propagation rates at the reference temperature (20 °C above the glass transition temperature in this case) the fracture surface energy is of the order of 10–100 kJ m^{-2}. The work of cohesion would be

Fig. 5.3. Variation of fracture surface energy with crack propagation rate. Butadiene/styrene rubbers: △ 96/4, □ 87.5/12.5. Butadiene/acrylonitrile rubbers: ● 82/18, ▲ 61/39. (After Gent, 1972, p. 326.)

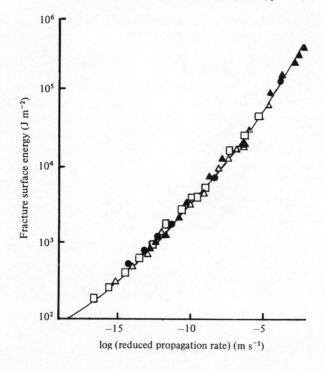

of the order of 0.1 J m^{-2} if calculated from an expression of the form of equation (5.1). However, in the case of a cross-linked elastomer the fracture process must involve the rupture of covalent bonds, and so Lake & Thomas (1967) calculated what should be the work of fracture if a more realistic molecular model involving such bond scission is used.

Lake & Thomas suggested that the forces in a rubber are transmitted primarily by the cross-links, and so all the bonds in the chain which connects two cross-links must be strained to virtually the same extent before one of the bonds breaks. Hence, if the energy required to rupture a monomer unit is u and there are on average $\bar{n}$ monomer units in the chain which connects two cross-links, then the energy required to break the chain is $\bar{n}u$, although only one of the units has been ruptured. Now the theory of rubber-like elasticity (Treloar, 1958, ch. 3) gives the mean distance between the ends of a chain which consists of $\bar{n}$ units each of length l as $\bar{L}$, where

$$\bar{L} = (8\bar{n}/3\pi)^{1/2} \, \gamma \, l, \tag{5.4}$$

and γ is a constant determined by the flexibility of the chain and usually having a value about 1.5. Now if N is the number of chains per unit volume, the number crossing unit area is given by $\frac{1}{2}\bar{L}N$, and so the fracture surface energy should be given by R, where

$$R = (\tfrac{1}{2}\bar{L}N)(\bar{n}u).$$

Hence, substituting from (5.4)

$$R = (2\bar{n}^3/3\pi)^{1/2} \, N\gamma lu, \tag{5.5}$$

and substitution of reasonable values in equation (5.5) gives a value for R of approximately 10–30 J m^{-2}.

It can be seen from figure 5.3 that, whilst the fracture surface energy for a rubber approaches this value at vanishingly low speeds of crack propagation, at experimentally accessible crack propagation rates the fracture surface energy is very much higher.

The reason for the very much greater rate of dissipation of energy at reasonable crack propagation rates is revealed by the very fact that time–temperature superposition on the basis of the WLF equation is applicable to the fracture surface energy. This suggests that the mechanism is related to the viscoelastic deformation of the rubber during the tearing process. During this process the rubber in the vicinity of the crack tip is subjected to a stress–strain cycle and will dissipate in unit volume an amount of energy given approximately by $\pi G'' \epsilon_{\max}^2$, where G'' is the loss modulus and $\epsilon_{\max}$ the maximum strain in the rubber. The fracture surface energy should, therefore, be proportional to the loss modulus of the rubber, and this is found to be the case as is indicated in figure 5.4 (Gent, 1972, p. 326).

It is thus suggested that the reason for the high fracture surface energy of rubbers is both that the 'intrinsic' fracture surface energy is high due to the straining of all the bonds in the chain at the crack tip, and that this intrinsic fracture surface energy is increased by the hysteresis losses as the crack propagates.

Generalised fracture mechanics

The role of hysteresis losses and, more particularly, the way in which these hysteresis losses appear in an expression for the fracture surface energy has been discussed by Andrews (1974) as one example of his 'generalised fracture mechanics'. Andrews' treatment goes far beyond crack propagation in rubbers, but a simplified approach to the generalised fracture mechanics treatment of rubber systems will be given here, and the extensions to other systems are dealt with in Andrews' publications (Andrews, 1974; Andrews & Billington, 1976; Andrews & Fukahori, 1977).

In order to examine the role of hysteresis losses, Andrews compares the energy made available at the crack tip as a result of crack propagation in a perfectly elastic material and in a material which suffers inelastic processes. He considers initially an infinite sheet of a reversibly elastic material which contains a crack of length $2a$. If this sheet is subjected to a stress σ_0 at infinity it demonstrates in the region away from the influence of the crack a strain ϵ_0 and a strain energy density W_0. The strain energy density W_P at a point P whose coordinates are (X, Y) referred to the centre of the crack as origin must, from dimensional considerations, be given by

$$W_P = W_0 \, f(X/a, Y/a, \epsilon_0). \tag{5.6}$$

Fig. 5.4. Relation between fracture surface energy and dynamic loss modulus G'' for styrene butadiene vulcanisates. Butadiene/styrene rubbers: △ 96/4, □ 87.5/12.5. Butadiene/acrylonitrile rubbers: ● 82/18, ▲ 61/39. (After Gent, 1972, p. 326.)

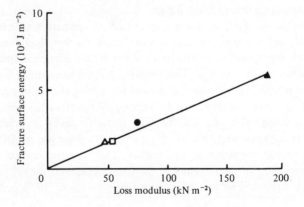

The strain ϵ_0 is included in equation (5.6) because for non-linear materials the distribution function f for the strain energy density is dependent upon the local strain, but this itself is fully characterised by its coordinates and ϵ_0. Andrews then introduces reduced coordinates $x = X/a$, $y = Y/a$, so that (5.6) may be rewritten

$$W_P = W_0 \, f(x,y,\epsilon_0). \tag{5.7}$$

When the crack extends, the strain energy density at P may increase or decrease, depending on the position of P. The change is given by

$$\frac{\mathrm{d}W_P}{\mathrm{d}a} = -\frac{W_0}{a}\left(x\frac{\partial f}{\partial x} + y\frac{\partial f}{\partial y}\right)$$

$$= -\frac{W_0}{a}\, g(x, y, \epsilon_0), \tag{5.8}$$

where g is another function of x, y and ϵ_0. Since the material is perfectly elastic, the function g applies whether at the point P the stress is increasing or decreasing.

The rate at which the total strain energy of the system changes with crack length is therefore given by $\mathrm{d}U/\mathrm{d}a$, where

$$\frac{\mathrm{d}U}{\mathrm{d}a} = \sum_P \frac{\mathrm{d}W_P}{\mathrm{d}a}\, \delta v; \tag{5.9}$$

δv is the volume element at P, and $\delta v = h\delta X\delta Y$, where h is the material thickness. Since $\delta v = ha^2\delta x\delta y$,

$$\frac{\mathrm{d}U}{\mathrm{d}a} = -W_0 \; ah \sum_P g(x, y, \epsilon_0)\, \delta x\delta y, \tag{5.10}$$

and so, since the crack area A is given by $A = 2ah$,

$$\mathscr{G} = -\frac{\mathrm{d}U}{\mathrm{d}A} = +\frac{W_0 a}{2} \sum_P g(x,y, \epsilon_0)\, \delta x\delta y.$$

Now, since all the terms being summed are dimensionless, the summation is independent of crack length and so

$$\mathscr{G} = k_1(\epsilon_0)aW_0, \tag{5.11}$$

where k_1 is a constant for a given ϵ_0.

It is now necessary to calculate the strain energy release rate for a system which displays hysteresis or other inelastic effects. For such a material it is not possible to carry out the summation indicated in equation (5.9), because the value of the strain energy density is not a unique function of the stress at that point. The strain energy density will depend upon whether the system is being loaded or unloaded, and if unloaded from what stress state it has been unloaded. Consequently, for an unloading step in an inelastic material, equation (5.8) must be written

$$\left(\frac{\mathrm{d}W_P}{\mathrm{d}a}\right)_{\mathrm{u}}^{\mathrm{i}} = -\frac{W_0}{a} \, G(x,y,\epsilon_0,\sigma_P),$$

where the superscript i refers to an inelastic material, the subscript u to an unloading process, and the function G is the function for an unloading process and must include the term σ_P representing the stress from which that element of material was unloaded. Hence, equation (5.10) for the whole body becomes

$$\left(\frac{\mathrm{d}U}{\mathrm{d}a}\right)^{\mathrm{i}} = -W_0\, ah \left[\, \sum_{P_1} g(x,y,\epsilon_0)\,\delta x \delta y + \sum_{P_{\mathrm{u}}} G(x,y,\epsilon_0,\sigma_P)\,\delta x \delta y \,\right],$$

(5.12)

where the summation over P_1 represents all points P which are loading and P_{u} represents all points P which are unloading from stress σ_P. Now if we write

$$\left(\frac{\mathrm{d}W_P}{\mathrm{d}a}\right)_{\mathrm{u}}^{\mathrm{i}} = \alpha\left(\frac{\mathrm{d}W_P}{\mathrm{d}a}\right)_{\mathrm{u}},$$

then α represents the ratio of the decrease in strain energy density for the unloading of an inelastic material to the decrease in strain energy density for the unloading of an elastic material for the same stress decrement. Thus α is a function of temperature, stress and rate of change of stress, i.e. $\alpha = \alpha(\sigma_P, T, \dot{a})$.

Since for a reversibly elastic system

$$\left(\frac{\mathrm{d}W_P}{\mathrm{d}a}\right)_{\mathrm{u}} = -\left(\frac{\mathrm{d}W_P}{\mathrm{d}a}\right)_{1},$$

equation (5.12) may be rewritten

$$\left(\frac{\mathrm{d}U}{\mathrm{d}a}\right)^{\mathrm{i}} = -W_0\, ah \left[\, \sum_{P_1} g(x,y,\epsilon_0)\delta x \delta y + \sum_{P_{\mathrm{u}}} \alpha g(x,y,\epsilon_0)\delta x \delta y \,\right. .$$

(5.13)

From (5.11)

$$\tfrac{1}{2} W_0 a \left[\, \sum_{P_1} g(x,y,\epsilon_0)\delta x \delta y + \sum_{P_{\mathrm{u}}} g(x,y,\epsilon_0)\delta x \delta y \,\right] = k_1(\epsilon_0) a W_0,$$

and so substituting in (5.13)

$$\left(\frac{\mathrm{d}U}{\mathrm{d}A}\right)^{\mathrm{i}} = a W_0 \left[k_1 - \sum_{P_{\mathrm{u}}} (1-\alpha) g(x,y,\epsilon_0)\delta x \delta y \right].$$

Now if $(1-\alpha)$ is put equal to β, then $\beta = \beta_0(\sigma_0, T, \dot{a}) g_1(x,y,\epsilon_0)$, where the function $g_1(x,y,\epsilon_0)$ is included to give the value of σ_P in terms of the remotely applied stress.

$$\left(\frac{\mathrm{d}U}{\mathrm{d}A}\right)^{\mathrm{i}} = a W_0 \left[k_1 - \beta_0 \sum_{P_{\mathrm{u}}} g_2(x,y,\epsilon_0)\delta x \delta y \right] = k_2(\sigma_0, T, \dot{a})\, a W_0, \quad (5.14)$$

where k_2 is another function.

$(dU/dA)^i$ is the actual release rate of energy which is available to form crack surfaces. Hence, if this is given the symbol $\mathscr{G}^i$, then for quasi-static crack propagation

$$\mathscr{G}^i = k_2 a W_0 = R,$$

but

$$\mathscr{G} = k_1 a W_0 = R_0,$$

where R is the apparent fracture surface energy which includes both the energy utilised at the surface and the energy dissipated by inelastic processes throughout the system, and R_0 is the energy which would be available at the crack surface if no inelastic processes occur. Hence,

$$\begin{aligned} R &= R_0(k_1/k_2) \\ &= R_0\ \Phi(\sigma_0,T,\dot{a}). \end{aligned} \tag{5.15}$$

There are a number of significant points concerning equation (5.15). The first is that it can be seen that the term involving the energy losses away from the actual site of bond scission acts as a multiplying term, rather than as an additive term to the energy losses at the crack tip. This is in complete contrast to the treatment put forward by Orowan for metals (see Knott, 1973, p. 110), in which the fracture surface energy may be written as $R = 2\gamma + p$, where p is the plastic work term.

The second major feature of equation (5.15) is that it suggests, contrary to the treatment of Orowan and others, that there is a direct relationship between the surface energy and the fracture surface energy, for if primary bonds are not disrupted during the propagation of the crack then R_0 is the same as $W_c = 2\gamma$. This point will be returned to in section 5.2.

5.2 Work of adhesion and the adhesive fracture surface energy
The work of adhesion

The expression which is analogous to equation (5.1) for an interfacial system is

$$(W_a)_{12} = (\pi/8)\, n_1 n_2\, (\epsilon_0)_{12}(r_0)^4_{12},$$

where the subscripts refer to the two phases 1 and 2, so that, in general, the theoretical reversible work of adhesion is of the same order of magnitude as that calculated previously for the work of cohesion. Mostovoy, Ripling & Bersch (1971) used a tapered double cantilever beam specimen with aluminium substrates and an epoxy adhesive to investigate the variation of fracture surface energy with crack propagation rate for a cross-linked brittle adhesive system. The results are shown in figure 5.5, and it can be seen that the adhesive fracture surface energy increases with crack propagation rate in the same manner as the cohesive fracture surface energy. It can also be seen that there is a strain energy release rate (in the case of the results shown in figure 5.5, about 45 J m^{-2}) below which the crack is unable to propagate at all. Ripling, Mostovoy & Bersch (1971) have termed this limiting value of the strain energy release rate $\mathscr{G}_{\text{Iscc}}$ by analogy with the stress corrosion cracking of metals where this term is used to describe the minimum value of the strain energy release rate below which crack propagation will not take place in the given corrosive environment. It may also be noted that this minimum value of the adhesive fracture surface energy is of the same order of magnitude as the minimum value of the cohesive fracture surface energy observed for elastomers. At the very slow strain rates, which correspond to the lowest observable crack propagation velocities, it is of course possible that the adhesive is behaving in a rubbery fashion.

Adhesive fracture surface energy for elastomeric systems

The adhesive fracture surface energy for systems involving elastomeric adhesives has been investigated by Andrews & Kinloch (1973a), and results for the adhesive joint formed between a styrene-butadiene-rubber (SBR) and a number of substrates are shown in figure 5.6 as a function of the crack propagation rate at the glass transition temperature, T_g (Andrews & Kinloch, 1973a). Included on the same graph is the variation of fracture surface energy with crack propagation rate, for a crack propagating entirely in the rubber. It can be seen that the energy dissipated in propagating a

Fig. 5.5. Variation of fracture surface energy with crack velocity for an adhesive system. Dow epoxy resin cured with varying amounts of tetraethylene pentamine (TEPA). ▲ 10 p.h.r. TEPA cured at 82 °C, □ 10 p.h.r. TEPA cured at 165 °C, ▽ 12.5 p.h.r. TEPA cured at 121 °C, × 12.5 p.h.r. TEPA cured at 101 °C, ● 15 p.h.r. TEPA cured at 65 °C, + 15 p.h.r. TEPA cured at 165 °C. (After Mostovoy *et al.*, 1971, p. 136.)

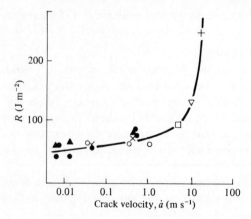

crack through the rubber was considerably greater than that which would have been necessary to cause propagation through the interface. This may be because in the former case rubber was being deformed and energy dissipated on both sides of the running crack, whereas, for a crack which propagated adjacent to a rigid substrate, energy could only have been dissipated on one side of the crack. It may also be because in the case of the cohesive fracture covalent bonds were being broken as the crack propagates.

The fact that all the lines in figure 5.6 are parallel to one another affords an interesting justification of the application of 'generalised fracture mechanics' to an adhesive system. Provided that in equation (5.15) the dependence of Φ on σ_0 is small, then equation (5.15) may be written for an adhesive system at a fixed temperature

$$R = R_0 \, F(\log r),$$

Fig. 5.6. Fracture surface energy as a function of crack propagation rate at T_g for five systems. A cohesive failure of styrene-butadiene-rubber, B adhesive failure of styrene-butadiene-rubber–FEPA (etched 120 s), C adhesive failure of styrene-butadiene-rubber–FEPA C20, D adhesive failure of styrene-butadiene-rubber–polyethylene terephthalate, E adhesive failure of styrene-butadiene-rubber–nylon 11. (After Andrews & Kinloch, 1973a, p. 393.)

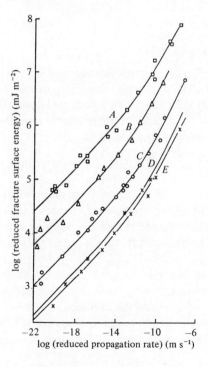

or

$$\log R = \log R_0 + F'(\log r),$$

where $F(\log r)$ is some function of the log of the crack propagation rate which is the same for all adhesive joints based on the same elastomer. Since at infinitely low crack propagation rates viscoelastic hysteresis losses should be zero, $F(\log r)$ must tend to unity as the crack propagation rate diminishes, and so R_0 is the reversible fracture surface energy. Since $F(\log r)$ is the same for all systems

$$\frac{(R)_{coh}}{(R)_{adh}} = \frac{(R_0)_{coh}}{(R_0)_{adh}} \tag{5.16}$$

where $(R)_{coh}$ and $(R)_{adh}$ are the fracture surface energies for a crack propagating cohesively as in graph A of figure 5.6 and adhesively as in graphs B to E. Since $(R_0)_{coh}$ has been determined independently by Lake & Lindley (1965), it is therefore possible to calculate $(R_0)_{adh}$ for various systems, and to compare this with the theoretical work of adhesion calculated from surface chemical measurements. This can be done by combining equation (3.2)

$$\gamma_{as} = \gamma_a + \gamma_s - 2(\gamma_a^d \gamma_s^d)^{1/2} - 2(\gamma_a^p \gamma_s^p)^{1/2} \tag{3.2}$$

and equation (2.1) written as

$$W_a = \gamma_a + \gamma_s - \gamma_{as}, \tag{2.1}$$

to give

$$W_a = 2(\gamma_a^d \gamma_s^d)^{1/2} + 2(\gamma_a^p \gamma_s^p)^{1/2}. \tag{5.17}$$

The values of γ_a^d, γ_s^d, γ_a^p and γ_s^p were calculated using a technique proposed by Kaelble (1970). The dispersion force and polar force contributions to the surface energy of a number of liquids are known (Fowkes, 1967; Dann, 1970), and so for a series of these liquids spreading on a given substrate it is possible to write equation (2.24') as

$$1 + \cos \alpha_i = \frac{2(\gamma_{li}^d \gamma_s^d)^{1/2}}{\gamma_{li}} + \frac{2(\gamma_{li}^p \gamma_s^p)^{1/2}}{\gamma_{li}} \quad (i = 1,2,3 \ldots), \tag{5.18}$$

where α_i is the contact angle made by liquid 1 on the solid surface s. The simultaneous solution of all the equations (5.18) yields values of γ_s^d and γ_s^p, and selected values are reported in table 5.1, together with W_a for the system SBR–polymeric substrate and R_0 for the system SBR–polymeric substrate.

In general it can be seen that there is good agreement between the observed 'intrinsic' fracture surface energy and the calculated work of adhesion. In the case of the adhesion of the SBR elastomer to an etched co-polymer of tetrafluoroethylene and hexafluoropropylene (FEPA) substrate it seems likely (Andrews & Kinloch, 1973b) that chemical bon-

Table 5.1. *The intrinsic fracture surface energy and the work of adhesion for SBR adhering to various substrates*

	γ_s^d (mJ m^{-2})	γ_s^p (mJ m^{-2})	γ_s (mJ m^{-2})	W_a SBR– polymer (mJ m^{-2})	R_0 SBR– polymer (mJ m^{-2})
SBR adhesive	27.8 ± 3.7	1.3 ± 1.1	29.1 ± 3.3		
FEPA	19.6 ± 3.7	0.4 ± 0.8	20.0 ± 2.9	48.4	21.9
PCTFE	31.4 ± 5.6	2.1 ± 1.5	33.5 ± 3.8	62.5	74.9
PET	41.8 ± 6.8	3.3 ± 2.8	45.1 ± 4.3	72.3	79.4
Nylon 11	43.1 ± 2.9	0.8 ± 0.6	44.0 ± 2.6	71.4	70.8
FEPA etched	34.4 ± 13.3	15.9 ± 13.4	50.3 ± 3.7	71.1	1780

(From Andrews & Kinloch, 1973a)
PCTFE: Polychlorotrifluoroethylene, PET: Polyethylene terephthalate.

ding occurs across the interface, leading to the rupture of primary bonds during fracture. In the case of the adhesion of the SBR elastomer to the unetched FEP substrate it is possible that poor wetting of the low-energy substrate may have led to a true area of contact which was less than the apparent area of contact.

Effect of environment on fracture surface energy

W_a as defined by equation (2.1) assumes that the two surfaces produced by cleavage of an interface are in contact with air or a vacuum. If, however, the interface is cleaved below the surface of a wetting liquid then the work of adhesion would have a different value, given by

$$(W_a)_l = \gamma_{al} + \gamma_{sl} - \gamma_{as},$$

where $(W_a)_l$ is the work of adhesion in the liquid l and γ_{al} and γ_{sl} are the adhesive-liquid and substrate-liquid interfacial energies, respectively. Hence, the change in the work of adhesion caused by immersion in a liquid l will be given (Gent & Schulz, 1972) by Δ, where

$$\Delta = W_a - (W_a)_l = \gamma_a + \gamma_s - \gamma_{al} - \gamma_{sl}.$$

So that if θ_{al} and θ_{sl} are the contact angles for the liquid on the adhesive and substrate, respectively, the introduction of Young's equation (2.17) yields

$$\Delta = \gamma_l(\cos \theta_{al} + \cos \theta_{sl}). \tag{5.19}$$

Gent & Schulz (1972) used this technique to measure the reduction in the work of adhesion between a styrene-butadiene-rubber and a sheet of Mylar (polyethylene terephthalate film) which could be brought about by immersion in a series of liquids including water. They also measured, by means of a peeling experiment, the fracture surface energy for this system immersed in the same series of liquids, and determined the ratio of the fracture surface energy in the given liquid, R_1, to the fracture surface energy in water, R_w, as a function of the reduction in the work of adhesion for that liquid, Δ.

Now from equation (5.16)

$$\frac{R_1}{R_w} = \frac{(R_0)_1 \, F \, (\log r)}{(R_0)_w \, F \, (\log r)} = \frac{(R_0)_1}{(R_0)_w} .$$

If we assume that under the circumstances of the experiment $(W_a)_1$ may be substituted for $(R_0)_1$ and $(W_a)_w$ for $(R_0)_w$, then

$$\frac{(R_0)_1}{(R_0)_w} = \frac{(W_a)_1}{(W_a)_w} = \frac{(W_a)_w - \Delta}{(W_a)_w} .$$

Hence

$$\frac{R_1}{R_w} = 1 - k\Delta. \tag{5.20}$$

The results of Gent & Schulz are shown in figure 5.7, and it can be seen that for many liquids the prediction of equation (5.20) is justified, i.e. the reduction in adhesive fracture surface energy (which from equation (4.17) is the same thing as the peel strength) may be attributed to the change in the work of adhesion. The results which fall well above the line in figure (5.7) were attributed to swelling of the rubber adhesive by the solvent.

Fig. 5.7. Fracture surface energy for joint rupture in the presence of various liquids, versus calculated reduction in work of adhesion. *A* water, *B* 10% ethanol, *C* 50% methanol, *D* ethanol, *E* butanol, *F* glycerol, *G* glycol, *H* formamide. (After Gent & Schulz, 1972, p. 288.)

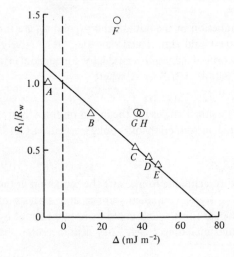

Gledhill & Kinloch (1974) attempted to apply a similar reasoning to the water induced failure of a mild-steel–epoxy joint. They calculated the work of adhesion in air, W_a, to be 291 mJ m^{-2} and the work of adhesion in water, $(W_a)_w$, to be -255 mJ m^{-2}, i.e. that the joint is unstable in water. On this basis Gledhill & Kinloch suggested that the joints should spontaneously debond in water, thus explaining the observed failure of such joints. Cherry & Thomson (1977), however, have pointed out that the total available energy which could cause the crack to propagate is 546 mJ m^{-2}, which is very much below the normally observed values for the strain energy release rate, below which crack propagation ceases. Consequently, Cherry & Thomson have suggested that the major cause of the environmental failure of some adhesive joints is not the thermodynamic instability of the joint, but the stresses induced at the interface as the polymeric adhesive solidifies. This will be dealt with further, below.

5.3 The strength of adhesive joints

The Griffith equation for an interface

If the adhesive fracture surface energy is known, then it should be possible, using the methods shown in chapter 4, to develop the relationship between the strength of an adhesive joint and the size of any flaws which may be present at the interface due to inadequate wetting of the substrate by the adhesive, or similar reasons. In fact, the necessary steps to develop the relationship analogous to equation (4.19) are a little more complicated. When a material is subjected to a tensile stress σ_1, then it contracts in the directions perpendicular to the direction of applied stress, and the strain in that direction is given by $\epsilon_2 = -\nu\sigma_1/E$, where ν is Poisson's ratio and E is Young's modulus. Consequently, if there is a crack at the interface between two materials and if a stress is applied perpendicular to that interface, then shear stresses as well as tensile stresses are developed at the interfaces, and more particularly at the tip of the crack in that interface. Rice & Sih (1965) have calculated an expression for the stresses at the tip of the crack taking account of these stresses, and Wang, Kwei & Zupko (1970) have shown that this expression is equivalent to the stress intensity factors, being given by

$$K_1 = \sigma(\text{sech } \pi\beta) (a)^{1/2}$$

and

$$K_2 = \sigma(2\beta \text{ sech } \pi\beta) (a)^{1/2} = 2\beta K_1$$

$$(5.21)$$

where K_2 is the stress intensity factor for forward shear mode cracking, and β is the so-called bi-material constant, defined by

$$\beta = \frac{1}{2\pi}\ln \frac{G_2\eta_1 + G_1}{G_1\eta_2 + G_2},$$

where G_1 and G_2 are the shear moduli of the two phases (note that $G = E/2(1 + \nu)$) and η_1 and η_2 are given by $\eta = (3 - \nu)/(1 + \nu)$.

Malyshev & Salganik (1965) similarly quoted an expression analogous to equation (4.10) for the relationship between the strain energy release rate and the stress intensity factor for a crack at an interface between dissimilar media. Their expression was

$$\mathscr{G} = \frac{(\pi/8)\,[G_1 + G_2(3 - 4\nu_1)]\,[G_2 + (3 - 4\nu_2)G_1]}{G_1 G_2\,[G_1(1 - \nu_2) + G_2(1 - \nu_1)]}\,(K_1^2 + K_2^2), \qquad (5.22)$$

so that equations (5.21) and (5.22) can be combined for the point of instability when $\mathscr{G} = \mathscr{G}_c = R$ and $\sigma = \sigma_f$ (the fracture strength of the joint), to yield an expression of the form

$$\sigma_f = \left(\frac{RQ}{a}\right)^{1/2}, \qquad (5.23)$$

where Q is a constant whose value can be calculated from equations (5.21) and (5.22).

Wang *et al.* (1970) fabricated a series of butt joints between sheets of aluminium and an epoxy adhesive, and created flaws of known length at the adhesive–adherend interface by incorporating short lengths of Teflon tape there, before curing the adhesive. They then tested the joint in uniaxial tension, and the results are shown in figure 5.8.

The first feature of figure 5.8 which should be noted is that the linear dependence of σ_f on $a^{1/2}$ is accurately obeyed. Wang *et al.* quote the physical properties of the adhesive and substrate as:

Fig. 5.8. Plot of fracture strength against half crack length. (After Wang *et al.*, 1970, p. 136.)

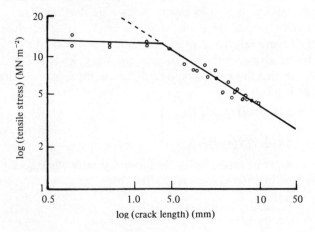

Material	Young's modulus (GPa)	Poisson's ratio
Epoxy	1.5	0.33
Aluminium	69	0.33

From equation (5.21) it is possible to calculate K_1 and K_2 from the value of the stress to break for a given crack length, and this gives $K_{1c} = 0.385$ MPa m$^{1/2}$ and $K_{2c} = 0.08$ MPa m$^{1/2}$. From equation (5.22), insertion of the appropriate figures yields

$$\mathcal{G} = 1.77 \times 10^{-9} \; (K_1^2 + K_2^2),$$

and so by combining these data we obtain $\mathcal{G}_c = R = 273$ J m^{-2}, which is close to the figures obtained by other workers for this type of system.

At crack lengths shorter than 2 mm, the stress required to break the joint is less than that predicted by equation (5.21). This may be due to the fact that stress concentrations associated with the edge of the joint are greater than those due to the presence of the flaw, and consequently the crack propagates from the edge of the specimen rather than from the end of the crack. It should, however, be noted that the ultimate tensile strength of the epoxy adhesive is 268 MPa. This implies that since $K_{1c} = 0.385$ MPa m$^{1/2}$, a flaw which has a half diameter less than 0.2 mm will be ineffective as a crack initiator since the system will fail under increasing load by rupture of the adhesive before the crack at the interface propagates.

Changes of failure mode in adhesive joints

Figure 5.9 (Cherry & Holmes, 1971) shows the variation in strength of a polyethylene–stainless-steel lap shear joint tested over a range of temperatures and strain rates. The dashed lines represent the case when failure was by brittle fracture initiated from a flaw at the interface, and the solid lines represent the cases when failure was by yield of the adhesive. It can be seen that a transition from yield behaviour to brittle fracture behaviour may be associated with either an increase in strain rate or a decrease in temperature. An explanation for the transition has been proposed which is based on the diagram shown in figure 5.10. Brittle fracture behaviour will be governed by an equation of the form of the equation (5.23), i.e. $\sigma^2 a = $ const, at a given temperature. This is illustrated diagrammatically by curve A in figure 5.10. At a given temperature and strain rate the yield strength of the bulk polymer will be independent of the size of any flaws, and may be represented by curve B. Consequently, at a temperature at which both interfacially initiated fracture and bulk yielding of the polymer occur, if the flaws are greater in length than a critical length, a_0, failure will be by fracture, whereas if the defects are smaller than a_0, failure will be by yield and in general the fracture strength of the joint will be less than the yield strength. At a higher temperature the yield strength curve may fall to the position marked *C* in figure 5.10, whilst the fracture curve, being much less sensitive to variation in temperature, may have moved only to *D*; the result is that the critical flaw size has now increased and only very rarely will a flaw be large enough to initiate fracture. Similarly, if the temperature is dropped or if the strain rate is increased considerably then the yield curve might be moved to *E*, at which point failure will be predominantly by fracture.

5.4 Stresses in adhesive joints
Engineering design with adhesive joints

In section 3.1 it was pointed out that the theoretical maximum strength of an adhesive joint which relied solely on dispersion force interactions would be in excess of 200 MN m^{-2}. In common with nearly all homogeneous systems, the stress required to break an adhesive joint is much less than the theoretical value, but because the stress has to be applied through a very much greater distance the actual energy dissipation during the failure process is much greater than the theoretical reversible work of adhesion or work of cohesion. Whilst the actual strength

Fig. 5.9. Strength versus crack length in the temperature range +50 °C to −70 °C. Points and error bars are omitted for clarity. Solid lines represent failure by yield, dashed lines represent failure by fracture. (From Cherry & Holmes, 1971, p. 88.)

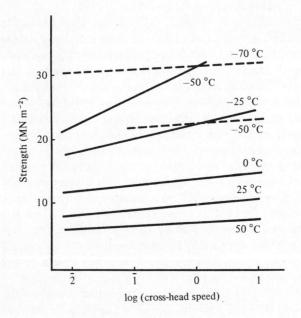

of an adhesive joint is controlled by precisely the same factors related to its defect structure as a homogeneous material, two additional features act to reduce the strength of adhesive joints even further, and so must be taken into account when designing an engineering structure. These will be termed the 'self-equilibrating stresses' and the 'service stresses'.

The term 'service stresses' will be used to describe the effect of stress concentrations which arise in an adhesive joint either due to its geometry or to the fact that there is a change in the elastic constants at the interface between adhesive and adherent so that additional stresses are set up there when a stress is applied to the system. 'Self-equilibrating stresses' in an adhesive joint are stresses, such as those occasioned by the shrinkage of one component relative to the other during the setting of the adhesive, which are present throughout the life of the joint and which add on to any applied stresses or any stresses which result from the application of a stress.

It may thus be seen that the actual stress which initiates failure is the sum of stresses arising from the application of an external force and of the self-equilibrating stresses, and may be very many times greater than the stress which might be calculated by the simple division of the applied force by the area of the joint. It may also be seen that the distinction between 'adhesive' and 'cohesive' failure which played such a major role in early

Fig. 5.10. Failure stress versus crack length. (From Cherry & Holmes, 1971, p. 91.)

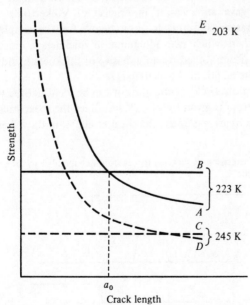

text-books on adhesion has very little foundation. True adhesive failure could only occur by fracture along the interface, and true cohesive failure could only occur by fracture in the bulk of the adhesive or by a yield process in which the bulk of the adhesive reached the yield point but not the region close to the interface. It seems better to use the term adhesive failure for those cases in which rupture takes place at or near an interface and is associated with stresses caused by the presence of the interface or with defects in the material arising from the proximity of the interface. The term cohesive failure may then be reserved for those cases in which failure is associated with stresses which reach a maximum in the bulk of the adhesive. A good example of the latter has been discussed by Gent & Lindley (1959).

Since the service behaviour of an adhesive joint is so dependent upon the service stresses and the self-equilibrating stresses present, the rest of this chapter will be devoted to a discussion of one particularly simple example of each type of stress: for any particular joint system the stress analysis must be carried out for that system if decisions are to be made as to how the strength of that system may be optimised.

Differential straining stresses

In a lap shear joint, stress concentrations arise due to the change in the strain in an adherend as the stress in the adherend is transferred into the adhesive and hence into the other adherend, as shown in figure 5.11. De Bruyne (1944) gave an account in English of Volkerson's earlier analysis of these stresses, and this is given below for the simple case shown in figure 5.12 in which two adherends of thickness s, breadth b, and tensile modulus E, are joined by an adhesive of thickness t, and shear modulus G, so that the length of the overlap is l.

Looking at the section $ABCD$ of the joint, it can be seen that the tensile strain in adherend 1 (ϵ_1) is given by $\epsilon_1 = P/sbE$. Since the shear stress, $d\tau$, acting on BB' is given by $d\tau = dP/b\,dx$, the shear strain, γ, in the adhesive is

Fig. 5.11. Transmission of shearing forces from one member to the other through the glue.

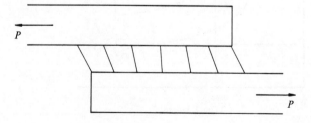

given by $\gamma = d\tau/G = dP/Gb\,dx$ and hence the displacement, δ, of B from its initial position is given by $\delta = t\,dP/Gb\,dx$. Now

$$P = \int_0^x dP = (Gb/t) \int_0^x \delta\,dx,$$ (5.24)

therefore, from the definition of ϵ_1

$$\epsilon_1 = (G/Est) \int_0^x \delta\,dx.$$ (5.25)

Now if the adhesive layer is thin, then the force transmitted across the interface between B and Q will be the same as that transmitted across the interface between C and R. Hence, the tensile strain in adherend 2 at CD (ϵ_2) is given by $\epsilon_2 = (X - P)/Ebs$. Therefore,

$$\epsilon_2 = \frac{X}{Ebs} - \frac{G}{Est} \int_0^x \delta\,dx.$$ (5.26)

From figure 5.12 it can be seen that

$$\delta = \delta_0 + \int_0^x \epsilon_1\,dx - \int_0^x \epsilon_2\,dx,$$

Hence from equations (5.25) and (5.26)

$$\delta = \delta_0 + \frac{G}{Est} \int_0^x \int_0^x \delta\,dx\,dx - \int_0^x \left(\frac{X}{Ebs} - \frac{G}{Est} \int_0^x \delta\,dx \right) dx$$

$$= \delta_0 - \int_0^x \frac{X\,dx}{Ebs} + \frac{2G}{Est} \int_0^x \int_0^x \delta\,dx\,dx.$$

From which

$$\frac{d^2\delta}{dx^2} = \frac{2G}{Est}\delta.$$ (5.27)

Fig. 5.12. Calculation of stress concentrations in a lap shear joint.

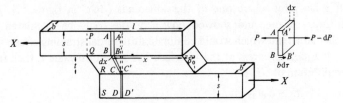

A solution of equation (5.27) is

$$\delta = A \cosh \lambda x + B \sinh \lambda x, \tag{5.28}$$

where $\lambda = (2G/Est)^{1/2}$, and A and B are determined from the conditions that when $x = 0$, $\delta = \delta_0$ and when $x = l$, $\delta = \delta_0$. This gives $A = \delta_0$ and $B = \delta_0(1 - \cosh \lambda l)/\sinh \lambda l$. Now

$$X = \int_0^l dP = (Gb/t) \int_0^l \delta \, dx.$$

Hence, from (5.28)

$$\delta_0 = \frac{X \lambda t \sinh \lambda l}{2Gb(\cosh \lambda l - 1)}.$$

The maximum shear stress is given $G\delta_0/t$, hence,

$$\tau_{\max} = \frac{X\lambda}{2b} \coth \frac{\lambda l}{2}.$$

It it is assumed that the joint will break when $\tau_{\max}$ reaches a critical value, τ_c, the load $X_{\max}$ which can be borne by the joint is given by

$$X_{\max} = \tau_c \, b \, l/p \coth p,$$

where

$$p = (Gl^2/2Est)^{1/2}.$$

For values of p greater than 2.7, $\coth p$ is within 1% of unity, i.e.

$$X_{\max} = b\tau_c \left(\frac{2Est}{G}\right)^{1/2}. \tag{5.29}$$

From equation (5.29) it can be seen that once a joint has reached a certain length, the strength is independent of the length of the overlap; increasing the length of overlap serves only to increase the stress concentration at the ends of the joint in proportion to the increase in overlap. The situation in a simple lap joint is far more complicated than has been indicated in the treatment above, but more detailed analyses are presented by Sneddon (1961).

Design to avoid differential straining stresses

A technique for designing the shape of an adhesive lap joint has been put forward by Cherry & Harrison (1970). They considered the case of a lap joint where one of the adherends (*AGC* in figure 5.13) has an arbitrary shape, and showed how a profile could be calculated for the other adherend which would eliminate differential straining stresses.

If the shear stress at the interface has a uniform value τ, then from the equilibrium of *GHA*

$$\tau x = E_1 t_{1x} \epsilon_{1x},$$

Fig. 5.13. Equilibrium of adherends of arbitrary shape to ensure uniform shear stress at the interface. (From Cherry & Harrison, 1970, p. 126.)

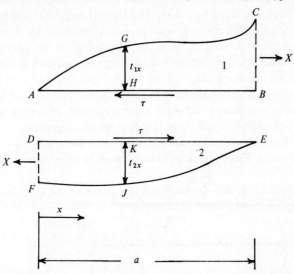

where E_1 and ϵ_{1x} are the modulus and the strain at section GH in adherend 1. Similarly

$$\tau(a - x) = E_2\, t_{2x}\, \epsilon_{2x}.$$

However, if the shear stress in the adhesive is to be constant throughout the adhesive, parallel lines before straining must remain parallel after straining, and so $\epsilon_{1x} = \epsilon_{2x}$, i.e.

$$E_1\, t_{1x}\, (a - x) = E_2 t_{2x}\, x. \tag{5.30}$$

Consequently, a joint designed according to (5.30) should be free of differential straining stresses, and figure 5.14 shows two joints designed on this principle.

Fig. 5.14. Lap shear joints designed to avoid differential straining stresses. (From Cherry & Harrison, 1970, p. 127.)

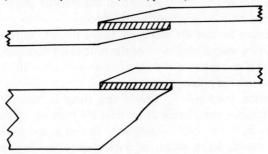

Self-equilibrating stresses

The effect of shrinkage stresses on the strength of an adhesive joint has been examined by Cherry and Hang (Hang, 1975). They formed joints between polycarbonate and an epoxy adhesive by allowing the adhesive to cure in contact with the polycarbonate to form a double cantilever beam fracture surface energy specimen, as shown in figure 5.15. The adhesive fracture surface energy for the system could then be determined by the techniques described in section 4.3.

In curing, the epoxy adhesive contracted and so set up shrinkage stresses at the adhesive–adherend interface whose magnitudes were determined by photoelastic measurements made when the system was examined in a polariscope. These shrinkage stresses could be reduced by compressing

Fig. 5.15. Double cantilever beam arrangement for the determination of the shrinkage strain energy release rate.

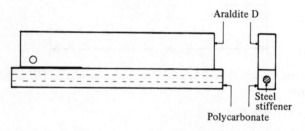

the polycarbonate specimen until its dimensions matched those which the epoxy would have assumed if it had been able to cure without the constraints imposed by its being bonded to the polycarbonate. The diminution of the shrinkage stresses was also monitored by photoelastic measurements.

The variation of the measured fracture surface energy with the compressive force applied to the polycarbonate substrate is shown in figure 5.16, in which it can be seen that as the compressive force increases, the fracture surface energy rises to a maximum which is nearly twice the initial value and then falls again. The maximum in the fracture surface energy curve corresponds with the minimum in the photoelastic stress pattern. At compressive stresses beyond the maximum in the fracture surface energy curve, the polycarbonate is shorter than the unstrained dimension of the epoxy resin and the relative shrinkage stresses increase with increasing compressive force on the polycarbonate.

The explanation proposed by Cherry and Hang is that the energy necessary to propagate a crack may derive not only from the applied stress system but also from the relief of shrinkage stresses as the crack propagates. In other words, the condition for crack propagation may be written

Fig. 5.16. Variation of apparent fracture toughness as shrinkage is removed by compression of substrate.

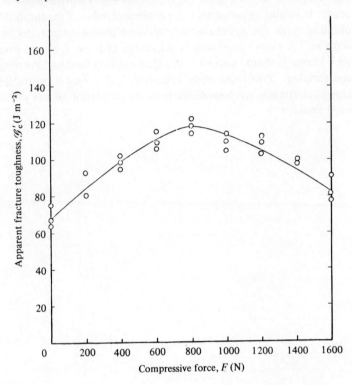

as

$$R = \mathscr{G}'_c + \mathscr{G}_s,$$

where R is the true fracture surface energy, $\mathscr{G}'_c$ is the measured critical strain energy release rate and $\mathscr{G}_s$ is the strain energy release rate associated with the relief of shrinkage stresses. It should be noted that since the crack propagates in mixed mode the value of R in the absence of compressive stress may be different from $\mathscr{G}_c$. However, it should also be noted that this value of R is the value which is normally determined and described as R_I. It can thus be seen that if a compressive force is applied to an adherend to counteract the effect of shrinkage stresses then $\mathscr{G}'_c$ must increase, and that if the shrinkage stresses are completely counteracted $\mathscr{G}'_c$ will reach a maximum value at which it equals R_I.

From figure 5.16 it can be seen that the difference between R_I and $\mathscr{G}_c$ is approximately 53 J m^{-2}, and so a conservative estimate of the strain energy release rate associated with the relief of shrinkage stresses, $\mathscr{G}_s$, is 53 J m^{-2}. When $\mathscr{G}_s$ was calculated from the magnitudes of the shrinkage stresses as determined by photoelastic measurements, a value of 62 J m^{-2} was obtained, showing good general agreement with the value obtained above. It should be noted that the maximum value of the shrinkage stress obtained with this system was determined photoelastically to be about 4MN m^{-2}, a stress comparable in magnitude with the measured strength of such joints. It should also be noted that epoxy adhesives have very much less shrinkage than most other adhesives, and it may be postulated that shrinkage stresses are responsible for a major portion of the weakness of most joints.

6 Friction and adhesion

6.1 The mechanisms of friction

When two polymers slide over each other, or when one polymer slides over a rigid substrate, the force of friction between them may be ascribed to two mechanisms. These are the adhesion mechanism and the deformation mechanism. At the simplest level, the adhesion mechanism arises from the interaction of intermolecular forces across the interface so that the work necessary to cause sliding of the surfaces over one another arises from the rupture of the intermolecular bonds of the type discussed in chapter 2. The deformation mechanism, on the other hand, arises from the mechanical interaction of the two surfaces, so that the work necessary to initiate or sustain sliding arises from the deformations in the surfaces caused by their mechanical contact. Obviously the distinction between the two mechanisms is somewhat arbitrary, since if adhesive bonds are formed across an interface then the surfaces will be deformed during the rupture process; however, in many cases it is found that friction due to adhesion can be differentiated from friction due to purely mechanical interactions. It is also possible to design experimental conditions so that only one mechanism of friction is acting at a given time; if two perfectly smooth surfaces are in contact, then the force required to initiate sliding must be due solely to adhesive interactions, whereas if a good lubricant is present, it prevents the contact of the surfaces and the development of interfacial bonds, and so only the deformation mechanism can operate. Since to a first approximation the two mechanisms are independent of one another, it is convenient to deal with them separately. This chapter will be concerned solely with the adhesion mechanism of friction.

6.2 The contact of solid surfaces

Unlike the liquid-liquid or the solid-liquid interface, the equilibrium configuration of a solid-solid interface is not controlled solely by surface forces, but is also governed by the mechanical properties of the materials on either side of the interface, since it is these mechanical properties which control the deformation of the solids. The exact nature of the equilibrium configuration of the interface will depend upon whether the deformations involved are elastic or plastic.

The elastic contact of solid surfaces

The derivation of quantitative expressions for the deformations and pressures involved when two elastic solids are pressed together by a force X is in general a lengthy procedure, and so it will not be repeated here as adequate treatments are available elsewhere (e.g. Timoshenko & Goodier 1970, p. 403 *et seq.*). The final results will, however, be quoted so that the particular expressions for specific configurations of the surfaces may be derived from them.

The general expression for the elastic contact of two spheres of radii R_1 and R_2, moduli E_1 and E_2, and Poisson's ratios ν_1 and ν_2 was developed by Hertz (1882). As shown in figure 6.1, contact is established on a plane, and if the radius of the circle of contact is given by a,

$$a^3 = \frac{3}{4}\left(\frac{1-\nu_1^2}{E_1} + \frac{1-\nu_2^2}{E_2}\right)\frac{R_1 R_2}{R_1 + R_2} X. \tag{6.1}$$

As a result of the local compression of the region of contact, distant points in the two spheres approach each other by a distance δ, where

$$\delta^3 = \frac{9}{16}\left(\frac{1-\nu_1^2}{E_1} + \frac{1-\nu_2^2}{E_2}\right)^2 \frac{R_1 + R_2}{R_1 R_2} X^2. \tag{6.2}$$

The distribution of pressure P across the circle of contact is given by

$$P = \frac{3X}{2\pi a^2}\left(1 - \frac{r^2}{a^2}\right)^{1/2}, \tag{6.3}$$

Fig. 6.1. The Hertzian contact of two elastic spheres.

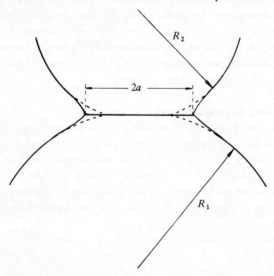

where r is the distance from the centre of the circle of contact.

The general equations (6.1) to (6.3) can be simplified for special cases when one of the surfaces is planar ($R = \infty$) or rigid ($E = \infty$). Then, for both the case of an elastic sphere in contact with a rigid planar surface and for a rigid sphere in contact with an elastic planar surface, the radius of the circle of contact, a, is given by

$$a^3 = \frac{3}{4}(1 - \nu^2)\frac{RX}{E}, \tag{6.4}$$

and distant points in the two materials approach each other a distance δ, given by

$$\delta^3 = \frac{9}{16}(1 - \nu^2)^2 \frac{X^2}{RE^2}, \tag{6.5}$$

from which it can be seen that

$$\delta = a^2/R. \tag{6.6}$$

The hardness of rubbers

The standard rubber hardness test consists essentially of the measurement of the increase in the depth of penetration of a rigid spherical indenter into a flat pad of rubber when an applied load is increased by a standard amount. The hardness is neither quoted as the increase in the depth of penetration nor even as the modulus of elasticity, but for the 'International Rubber Hardness' scale as an index which varies from 0 for an infinitely large to 100 for an infinitely small increase in penetration. The important thing to note, however, is that the hardness of a rubber measures its modulus of elasticity. The relationship between the various measures of rubber hardness and the modulus of elasticity has been discussed in some detail by Gent (1958). Figure 6.2 shows the indentation of a planar rubber surface by a spherical indenter as in a standard rubber hardness test. From the geometric theorem which relates the areas on intersecting chords of a circle, $H = a^2/2R$, and so it can be seen that this plane lies a distance $\frac{1}{2}\delta$ below the surface of the undeformed rubber.

Fig. 6.2. Indentation of an elastic half-plane by a rigid spherical indenter.

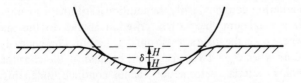

Indentation by a punch

The indentation of a planar elastic surface by a rigid planar punch has been discussed by Lur'e, and for the case of a plane-ended circular punch (Lur'e, 1964, p. 273), provided that the line of action of the applied force lies within a cylinder of radius $a^1/3$, where a^1 is the radius of the punch, then contact is established over the entire face of the punch, i.e. $a_0 = a^1$. The pressure at any point on the area of contact is given by

$$P = \frac{X}{\pi a^2} \; \frac{1}{2(1 - r^2/a^2)^{1/2}},$$ (6.7)

and the total indentation is given by δ, where

$$\delta = \frac{1 - \nu^2}{2} \frac{X}{Ea},$$ (6.8)

It can be seen that the pressure at the edges of a circular punch would rise to infinite values according to equation (6.7). In fact, before this can happen the yield point of the material is exceeded and plastic flow obtains.

The plastic contact of solid surfaces

When the contact pressure between two surfaces is high enough, one or both of the contacting bodies will yield. For many polymers the nature of the yield point is ill-defined. However, a treatment of plastic contact in which one of the bodies behaves as an elastic/plastic solid has been derived by Hill (1960, p. 254), and the final results will be quoted below as they give a valuable insight into the form of the behaviour of polymer surfaces.

If the case of the contact of a rigid sphere with a softer plane is considered, the initial contact will be elastic and can be described by equations (6.1) to (6.3). The elastic analysis (Timoshenko & Goodier, 1970) indicates that the maximum shear stress, τ_m, in the system occurs at a point approximately $0.5a$ below the centre of the circle of contact and has a value of approximately $0.46P_a$ where $P_a = X/\pi a^2$ and is the mean pressure over the contact area. If the polymer yields when the maximum shear stress reaches a critical value τ_y, then as soon as $0.46P_a$ is equal to τ_y the material will start to yield at that point. In fact, a pressure-dependent von Mises yield criterion is probably a better criterion, but the difference between the results obtained on that basis and on the basis of a maximum shear stress criterion does not justify the additional arithmetic manipulation involved. If a maximum shear stress criterion applies to the material, $\tau_y = \frac{1}{2}\sigma_y$, where σ_y is the yield stress in uniaxial tension, consequently yield starts to occur when $P_a = 1.1\sigma_y$. However, because of the constraints of the surrounding material, deformation cannot continue under this stress and it is found that the mean stress has to rise to approximately $2.97\sigma_y$ (Hill, 1960, p. 260) before all the material beneath the indenter is in a fully plastic condition. Cottrell (1964, p. 327) shows that for a flat cylindrical punch P_a must reach approximately $2.82\sigma_y$ for a non-work-hardening material before all the material beneath the punch is in a fully plastic condition.

The hardness of plastic surfaces
Various instruments have been designed to measure the 'hardness' of a plastic surface by measuring the depth of indentation left when a hemispherically ended rigid indenter is pressed into a planar surface and then removed. In this case it is the yield point of the material rather than the elastic modulus which is being measured. If the 'hardness' of a plastic is defined as the average pressure on an indenter when it is pressed into a surface, it can be seen that if this is given the symbol h then

$$h \approx 3\sigma_y.$$

It can also be seen that there is a comparatively narrow range of contact pressures over which the nature of the deformation changes from one that is predominantly elastic to one that is predominantly plastic.

The adhesion of contacting surfaces
Equation (6.1) was derived without reference to surface forces. In general this neglect is justified since the increase in surface area due to surface forces is negligible. For low modulus materials, however, surface forces may cause a detectable increase in the area of contact, and the work of adhesion can be determined from this increase.

Johnson, Kendall & Roberts (1971) determined the area of contact of two spheres when pressed together by a force X. They used the condition that at equilibrium the total free energy of the system should be a minimum and considered the contribution to the total free energy of the system of the stored elastic energy U_E, the mechanical potential energy of the deforming forces U_M, and the surface energy U_S.

Under Hertzian conditions, the radius of the circle of contact is given by (from equation (6.1))

$$a^3 = RX/K, \tag{6.1'}$$

hence from (6.6)

$$\delta = R^{-1/3} X^{2/3} K^{-2/3}, \tag{6.6'}$$

where $R = R_1 R_2/(R_1 + R_2)$ and $1/K = \frac{3}{4}[(1 - \nu_1^2)/E_1 + (1 - \nu_2^2)/E_2]$. However, in a real system, owing to the action of surface forces, the actual radius of contact a_1 brought about by the application of a force X_0 is greater than a_0 ($a_0^3 = RX_0/K$). Johnson *et al.* therefore suggested that the total stored elastic energy could be calculated by determining the work done when, first of all the two spheres are pressed together by a force X_1 (= Ka_1^3/R) and then secondly the system is unloaded from X_1 to X_0 maintaining contact over the whole of the circle of contact of radius a_1. The second step corresponds to the unloading of a punch indenter for which the force-deflection relationship is given by equation (6.8). The complete cycle of operations is shown in figure 6.3.

If the energy changes involved in the two steps are U_1 and U_2, then

$$U_E = U_1 + U_2$$

where

$$U_1 = \int_0^{\delta_1} X \mathrm{d}\delta = \int_0^{X_1} \frac{2}{3} \frac{X^{2/3}}{K^{2/3}R^{1/3}} \, \mathrm{d}X \quad \text{(from equations (6.1') and (6.6'))}$$

Fig. 6.3. (a) The contact between two smooth elastic solids: a_0 represents the contact radius in the absence of surface forces, and a_1 the contact in the presence of surface forces. (b) The hypothetical loading situation which enables a_1 to be calculated. (After Johnson *et al.*, 1971, p. 303.)

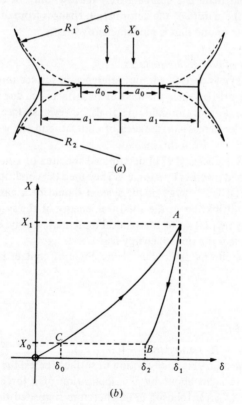

(a)

(b)

i.e.

$$U_1 = \frac{2}{5} \frac{X_1^{5/3}}{K^{2/3}R^{1/3}}$$

and

$$U_2 = \int_{\delta_1}^{\delta_2} X \mathrm{d}\delta = \int_{X_1}^{X_0} \frac{2}{3} \frac{X}{Ka_1} \, \mathrm{d}X \quad \text{(from equation (6.8))}$$

$$= \frac{1}{3K^{2/3}R^{1/3}} \left(\frac{X_0^2 - X_1^2}{X_1^{1/3}} \right),$$

i.e.

$$U_E = \frac{1}{K^{2/3}R^{1/3}} \left(\frac{X_1^{5/3}}{15} + \frac{X_0^2}{3X_1^{1/3}} \right). \tag{6.9}$$

The mechanical potential energy of the system, U_M, is given by the work done by externally applied forces, i.e.

$$U_M = -X_0 \delta_2 = -X_0 [\delta_1 - (\delta_1 - \delta_2)]$$

$$= -X_0 \left(\delta_1 - \frac{2}{3} \frac{(X_1 - X_0)}{Ka_1} \right)$$

$$= -X_0 \left(\frac{X_1^{2/3}}{K^{2/3}R^{1/3}} - \frac{2}{3} \frac{K^{1/3}}{R^{1/3}X_1^{1/3}} \frac{(X_1 - X_0)}{K} \right),$$

i.e.

$$U_M = -\frac{X_0}{K^{2/3}R^{1/3}} \left(\frac{X_1^{2/3}}{3} + \frac{2X_0 X_1^{-1/3}}{3} \right). \tag{6.10}$$

The surface energy U_S is given by $\pi a_1^2 W_a$, where W_a is the work of adhesion, i.e.

$$U_S = \pi W_a (RX_1/K)^{2/3}. \tag{6.11}$$

The total energy of the system is then given by

$$U_T = U_S + U_M + U_E,$$

and the condition for equilibrium is that $\mathrm{d}U_T/\mathrm{d}\delta = 0$, which may be written as $\mathrm{d}U_T/\mathrm{d}X_1 = 0$. Hence, from (6.9), (6.10) and (6.11)

$$\frac{\mathrm{d}U_T}{\mathrm{d}\delta} = \frac{1}{K^{2/3}R^{1/3}} \left(\frac{X_1^{2/3}}{9} - \frac{X_0^2 X_1^{-4/3}}{9} - \frac{2X_1^{-1/3}X_0}{9} + \frac{2X_0^2 X_1^{-4/3}}{9} \right)$$

$$- \frac{2\pi W_a R^{2/3} X^{-1/3}}{3K^{2/3}}$$

$$= \frac{X_1^{-4/3}}{9K^{2/3}R^{1/3}} (X_1^2 - X_0^2 - 2X_1 X_0 + 2X_0^2 - 6\pi W_a RX_1) = 0.$$

Solving this equation for X_1 gives, taking the positive root,

$$X_1 = X_0 + 3\pi W_a R + [6\pi W_a RX_0 + (3W_a\pi R)^2]^{1/2}. \tag{6.12}$$

Now, from equation (6.1), $a_1^3 = RX_1/K$, and so a_1 can only reduce to zero if X_0 is negative. If X_0 is negative, a real solution to (6.12) only exists for the case

$$6\pi W_a RX_0 \leqslant (3W_a\pi R)^2,$$

where the equality represents the limiting case of the two spheres just touching. Consequently, it may be seen that a force of $\frac{3}{2}W_a\pi R$ is necessary to separate the spheres. This force represents the adhesion of the two solid surfaces. If a value for W_a of 70 mJ m^{-2} is assumed (as would appear reasonable for two rubber surfaces), the adhesive force for 5 cm radius spheres is approximately 16×10^{-3} N.

Johnson *et al.* (1971) measured the area of contact of optically smooth rubber balls as a function of the load which pressed them together. These experiments were carried out with dry surfaces, with surfaces immersed in water and surfaces immersed in a dilute solution of detergent, and the results are shown in figure 6.4. It was found that the curves could be fitted well to equation (6.12) if values for W_a of 71 ± 4 mJ m^{-2} were taken for the interface in air, of 6.8 ± 0.4 mJ m^{-2} for the interface in water, and zero for the detergent-treated interface. Since the first two results imply

Fig. 6.4. Variation of contact area for two rubber spheres ($R_1 = R_2 = 2.2$ cm) in dry and lubricated contacts. + dry contact, o water contact, × soap solution contact, --- Hertz theory, – modified theory. (From Johnson *et al.*, 1971, p. 308.)

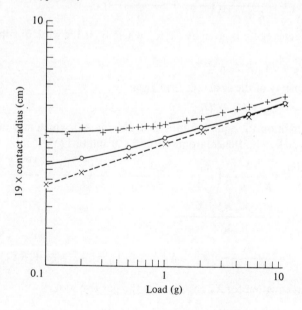

that for the rubber surface $\gamma_s = 35$ mJ m^{-2} and for the rubber–water interface $\gamma_{sl} = 3.4$ mJ m^{-2}, the contact angle for water on rubber can be predicted using Young's equation (2.17), and the known surface tension of water, 72.5 mJ m^{-2}. From this $\theta = 64°$, which agrees well with the experimentally determined value of 66°.

6.3 The adhesion theory of friction
Amontons' laws

Since solids adhere to each other, it is not surprising that a frictional force is observed when an attempt is made to slide one body over another. However, any quantitative treatment has to explain the basic laws of friction which were presented by Amontons as early as 1699 as:

(i) The frictional force between two bodies is independent of the area of contact.

(ii) The frictional force between two bodies is directly proportional to the load normal to the interface.

In order to give a basis for Amontons' laws, Bowden & Tabor (1950, 1964) pointed out that few, if any, real solids are smooth on the microscopic scale and consequently that the true area of contact is confined to that comparatively small proportion of the apparent area of contact where asperities in the two surfaces meet, as suggested in figure 6.5. At these points true interfacial contact is made. Motion of the surfaces over each other is only possible if the interface or the material near to it is sheared. The frictional force due to adhesion is then given by

$$F_a = \tau_s \Omega, \tag{6.13}$$

where τ_s is the shear strength of the interface and Ω is the true area of contact of the two surfaces. The variation of the frictional force with such factors as the load normal to the interface, the hydrostatic stress, the velocity of sliding, the temperature, etc., then resolves itself into a discussion of the effect of these factors on either τ_s or Ω. If the deformation of the asperities is fully plastic, then it can be seen from the discussion of 'hardness' for a plastically deforming material that if the hardness is h, then the area of contact due to one asperity will be given by $\Omega_1 = X_1/h$, where X_1 is the load borne by that asperity. Hence, for an assembly of asperities which make up the contacting area

Fig. 6.5. Contact of two rough surfaces.

$$\Omega = \Omega_1 + \Omega_2 + \Omega_3 + \ldots = \frac{X_1}{h} + \frac{X_2}{h} + \ldots = \frac{X}{h}. \tag{6.14}$$

Consequently, from (6.13)

$$F_a = X \tau_s/h = \mu X, \tag{6.15}$$

where μ is the coefficient of friction, which equals τ_s/h.

Equation (6.15) provides an explanation for Amontons' second law and also serves as a definition for the coefficient of friction. However, since $h = 3\sigma_y = 6\tau_s$, if a maximum shear stress criterion for yield is assumed, this also suggests that μ should have a value of 0.18.

Polymers do not in general obey Amontons' laws to anything better than a first approximation. A typical variation of frictional force with applied load is shown in figure 6.6 for polytetrafluoroethylene (Lancaster, 1972, quoting Allen, 1958), from which it can be seen that over a wide range of loads the variation of friction with load is represented by

$$F = k X^n,$$

or

$$\mu = \frac{F}{X} = kX^{n-1}. \tag{6.16}$$

Pascoe & Tabor (1956) have shown that for the static friction between specimens of the same polymer, for many polymers $(n - 1)$ lies between -0.2 and -0.30, typical results being shown in table 6.1.

In general it can be seen that the coefficient of friction decreases with an increase in the normal force. From equation (6.15), $\mu = \tau_s\Omega/X$. Consequently, for plastic deformation the coefficient of friction can only decrease if the true area of contact is not accurately proportional to the applied normal force since, although the shear strength of the interface may be a function of the true normal force, the prerequisite for a change in the true normal force is a variation in the true contact area which is not proportional to the applied normal force. It is therefore necessary to examine how the true area of contact between two rough solids varies with the applied load.

The true area of contact between rough solids

Greenwood & Williamson (1966) have examined the problem of the true area of contact between a rough surface with N_0 asperities per unit area and a flat surface. (In a later paper Greenwood (1967) shows that his results are not materially altered if one considers the contact of two asperity-bearing surfaces.) They assume that the probability of a given asperity having a height z above a reference plane in the surface is some function of z, which they call $\phi(z)$. If the flat surface approaches within a distance d of a reference plane in the contacting surface, then the total

Table 6.1. *Value of exponent (n − 1) in expression* $\mu = kX^{n-1}$ *for various polymers*

Polymer	$(n-1)$
Polyvinylidene chloride	−0.17
Polytetrafluoroethylene	−0.2
Polymethyl methacrylate	−0.23
Nylon 66	−0.26
Polyethylene terephthalate	−0.26
Polyethylene (branched)	−0.26

number of contacts is $\int_d^\infty N_0\phi(z)\mathrm{d}z$. In general neither the asperity shape, nor the force-deflection relationship for an individual asperity are known; however, both the load borne by a single asperity and the area of contact due to that asperity will be functions of the amount by which the asperity has been squashed, i.e. $(z - d)$. Thus if the load borne by an asperity squashed by an amount $(z - d)$ is a function of $(z - d)$, $f(z - d)$, and if the true area of contact is another function, $g(z - d)$, then the expressions for the total load and total contact area are

Fig. 6.6. Variation of frictional force and coefficient of friction with load. (From Lancaster, 1972.)

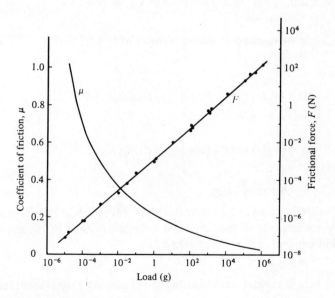

$$X = \int_d^\infty f(z - d) \, N_0 \, \phi(z) \mathrm{d}z, \tag{6.17}$$

$$\Omega = \int_d^\infty g(z - d) \, N_0 \, \phi(z) \mathrm{d}z, \tag{6.18}$$

Now in general $\phi(z)$ is not known, but if it follows a Gaussian distribution, $\phi(z) = 1/[\sigma(2\pi)^{1/2}] \exp(-z^2/2\sigma^2)$, where σ is the standard deviation for asperity heights, then the upper ranges of this distribution (which are the only asperities involved in light contacts at least) may be approximated by the exponential distribution $\phi(z) = (1/\sigma)\exp(-z/\sigma)$. If this distribution is now introduced into (6.17) and (6.18) it is found that

$$X = N_0 \int_d^\infty (1/\sigma) \exp(-z/\sigma) f(z - d) \, \mathrm{d}z,$$

$$\Omega = N_0 \int_d^\infty (1/\sigma) \exp(-z/\sigma) g(z - d) \, \mathrm{d}z,$$

$$N = N_0 \int_d^\infty (1/\sigma) \exp(-z/\sigma) \, \mathrm{d}z,$$

where N is the total number of contacts. Or, writing $t = z - d$,

$$X = (N_0/\sigma) \exp(-d/\sigma) \int_0^\infty \exp(-t/\sigma) f(t) \, \mathrm{d}t, \tag{6.19}$$

$$\Omega = (N_0/\sigma) \exp(-d/\sigma) \int_0^\infty \exp(-t/\sigma) g(t) \, \mathrm{d}t, \tag{6.20}$$

$$N = (N_0/\sigma) \exp(-d/\sigma) \int_0^\infty \exp(-t/\sigma) \, \mathrm{d}t,$$

$$= N_0 \exp(-d/\sigma). \tag{6.21}$$

The implications of equations (6.19), (6.20) and (6.21) are far-reaching. Since the integrals on the right-hand sides of equations (6.19) and (6.20) are independent of d, it can be seen that

$$\Omega = \mathrm{const} \times X,$$

i.e. if there is an exponential distribution of asperity heights then, independently of the mode of deformation of the asperities (i.e. whether it is elastic, plastic or something in-between) or the shape of the asperities, there is exact proportionality between the true area of contact and the load. This provides a basis for Amontons' laws irrespective of the nature of the material.

It can also be seen that the mean pressure $P_a = X/\Omega$ is a constant, and that the load borne by a given asperity, i.e. X/N, is a constant irrespective of the applied load, as long as the exponential distribution of asperity heights remains. It is to be expected that such relationships will break down when local true pressures reach such a magnitude that the exponential distribution of asperity heights no longer obtains.

The elastic–plastic transition

From the discussion of plastic deformation in section 6.2, it can be seen that the onset of plastic flow occurs when the mean pressure over an asperity is given by $P_a = 1.1\sigma_y$, whereas $h \approx 2.82\sigma_y$. Yielding commences therefore at a mean pressure of $0.39h$. The mean pressure is given from equation (6.4) as

$$P_a = \frac{X}{\pi a^2} = \frac{X}{\pi R \delta} = \frac{4Ea}{3\pi R (1 - \nu^2)} ,$$

and so substituting from (6.6) it can be seen that the value of δ at which yielding commences is given by δ_p where

$$\delta_p = \frac{9\pi^2 (0.39)^2 h^2 R (1 - \nu^2)^2}{16E^2} \tag{6.22}$$

$$\approx R h^2 / E'^2,$$

where $E' = E/(1 - \nu^2)$. If the distribution of asperities follows an exponential pattern, then the probability of an asperity making contact with the other surface is given by

$$P(z > d) = \int_d^\infty (1/\sigma) \exp(-z/\sigma)\, dz.$$

Similarly, the probability that the contact is plastic is given by

$$P(z > (d + \delta_p)) = \int_{d + \delta_p}^\infty (1/\sigma) \exp(-z/\sigma)\, dz.$$

Hence, by the reasoning that leads to equation (6.20), the ratio of the area of plastic contact to the total area of contact is given by

$$\frac{\Omega_p}{\Omega} = \frac{\exp[-(d + \delta_p)/\sigma]}{\exp(-d/\sigma)} = \exp(-\delta_p/\sigma),$$

which from equation (6.22) gives

$$\frac{\Omega_p}{\Omega} \approx \exp\left(-\frac{R}{\sigma}\frac{h^2}{E'^2}\right) .\tag{6.23}$$

Equation (6.23) demonstrates the initially quite surprising fact that the nature of a contact is not a function of the applied load, but solely a function of the material properties of the solids and the surface topography.

Greenwood & Williamson (1966) used the term 'plasticity index', ψ, where

$$\psi = \frac{E'}{h}\left(\frac{\sigma}{R}\right)^{1/2},\tag{6.24}$$

to describe the potential for forming plastic contacts. From (6.23) and (6.24) it can be seen that

$$\frac{\Omega_p}{\Omega} - \exp\left(-1/\psi^2\right),$$

and so for $\psi < 0.5$, $\Omega_p/\Omega < 0.18$, whereas for $\psi > 1.5$, $\Omega_p/\Omega > 0.65$, hence for surfaces for which $\psi > 1.5$ the contact is predominantly plastic, whereas for $\psi < 0.5$ the contact is predominantly elastic. Using values of $h = 3\sigma_y$, values of E'/h are reported for a number of metals and plastics in table 6.2.

Greenwood & Williamson (1966) have shown that $(\sigma/R)^{1/2}$ for a number of metal surfaces varies between 3×10^{-1} for a rough bead blasted surface and 4×10^{-3} for a highly polished surface, hence it can be seen that polymers are very much more likely to form elastic contacts at the asperities than are metals, and that a low-modulus polymer such as low-density polyethylene probably never forms fully plastic contacts at the tips of the asperities, irrespective of what happens to the bulk deformation of the solid. Obviously some plastic flow must occur since the mean pressure must rise above $1.1\sigma_y$, but the majority of the deformation must be elastic,

Table 6.2. *Values of E'/h for some materials*

Material	E'/h
Low-density polyethylene	8
Polytetrafluoroethylene	9
High-density polyethylene	12
Nylon 66'	17
Polymethyl methacrylate	21
Epoxy resin	26
Phenol formaldehyde resin	83
Aluminium	110
Steel	200

and hence if the indenter is removed the asperities still remain and have not been flattened.

Junction growth

Although the nature of the contact, whether it is elastic or plastic, does not affect the true area of contact under the action of a normal load, it may give rise to variations in behaviour when a load is applied which is tangential to the interface. Sliding can not commence until the interface, or the material close to it, yields, and this implies that the applied stresses have satisfied the appropriate yield criterion. For most plastics a von Mises yield criterion which is dependent upon the hydrostatic pressure appears to be appropriate (Williams, 1973, p. 70) but, neglecting the pressure dependence, an approximation in two dimensions for the system shown in figure 6.7 may be written (Bowden & Tabor, 1964, p. 72) as

$$\sigma^2 + \alpha\tau^2 = k, \tag{6.25}$$

where σ and τ are the normal and tangential stresses and k and α are constants. If the junction has already yielded before any tangential stress is applied (i.e. plastic contact) then $\sigma^2 = k$. As soon as a tangential force is applied then, however small this force is, $\sigma^2 + \alpha\tau^2 > k$ and the system must yield under the action of the normal force so that the contact area increases until σ is reduced to a lower value appropriate to equation (6.25). However, any further increase in τ would cause a further increase in contact area and there is nothing in the model so far proposed which indicates how the process could stop. Junction growth, as the above process is called, undoubtedly takes place in metals, where it is eventually limited by the presence of weaker surface layers, work hardening of the junctions and other reasons (Bowden & Tabor, 1964, p. 74 *et seq.*). It has been suggested that junction growth does not occur with polymers (Pascoe & Tabor, 1956), and a possible reason is that the polymers which were

Fig. 6.7. Junction growth under the action of a tangential stress.

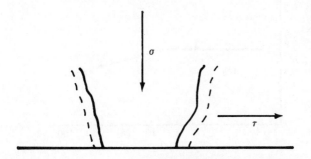

examined were those in the upper part of table 6.2 for which the contacts were predominantly elastic so that the effect of normal stress would be to reduce slightly the tangential stress which caused yielding rather than to cause junction growth. It seems quite likely, however, that junction growth could occur with those polymers which have a high plasticity index.

The variation of μ with load

The reason for the decrease in the coefficient of friction with increasing load remains unresolved; however, in view of the results above it appears that an explanation should be sought in terms of elastic deformation of the surfaces rather than plastic deformation of the surfaces. Archard (1957) measured the coefficient of friction for crossed cylinders of Perspex (polymethyl methacrylate) as a function of the load, for three different surface finishes. The results are shown in figure 6.8 (Lancaster, 1972). It should be noted that the experimental conditions come very close to single point contact and that the mean pressure at the point of contact would be very high. Consequently, the explanation which may be offered for these results is that at low loads the distribution of asperity heights for the specimens of figures 6.8(*c*) and 6.8(*b*) is such that the true

Fig. 6.8. Variation of coefficient of friction with load for crossed cylinders of polymethyl methacrylate: (*a*) smooth, polished surfaces, (*b*) abraded surfaces, (*c*) turned surfaces. (From Lancaster, 1972.)

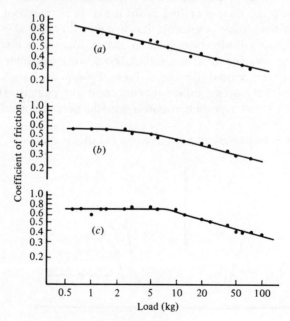

area of contact is proportional to the applied load, as expected from equations (6.19) and (6.20), and Amontons' laws are obeyed. At higher loads, however, the asperities are flattened and thus the true area of contact is equal to the apparent area of contact. If the deformation is Hertzian, then from equations (6.4) and (6.6)

$$\Omega \propto X^{2/3}, \tag{6.26}$$

or

$$\mu = \frac{\Omega}{X} = kX^{-1/3}, \tag{6.27}$$

where k is a constant. Hence it can be seen that for single point plastic contact $\mu = $ const $\times$ X, and for single point elastic contact $\mu = $ const $\times$ $X^{-0.33}$. Pascoe & Tabor (1956) have suggested that in view of the fact that polymers deform viscoelastically a more appropriate form of equation (6.26) for polymers would be

$$\Omega \propto X^{2/m},$$

where $2 < m < 3$; the condition $m = 2$ corresponds to a purely plastic contact, $m = 3$ corresponds to a purely elastic contact. Equation (6.27) can then be written as

$$\mu = kX^{(2-m)/m},$$

and $(2 - m)/m$ can be identified with $(n - 1)$ of equation (6.16) and table 6.1. For the highly polished specimen whose results are shown in figure 6.8(*a*), it is suggested that single point contact obtains from the lowest load and so no transition is observed.

It should be noted that the transition point from the region where Amontons' laws are obeyed to a region of diminishing coefficient of friction is not to be associated with a transition from elastic to plastic contacts, but just the point at which the surface asperities are sufficiently flattened, elastically, to give effectively a single point contact.

6.4 Adhesion and sliding friction

The discussion so far has been concerned with static friction, however, adhesive bonds are also formed during the sliding of two surfaces over each other and friction arises from the force required to disrupt the bonded surfaces. In chapter 5 it was suggested that whereas the energy required to propagate a crack through a glassy material is mainly dissipated by processes giving rise to plastic deformation, the energy required to propagate a crack through a rubber is dissipated mainly as hysteresis losses. Consequently, in the discussion of sliding friction, elastomers and plastics below their T_g will be considered separately.

The sliding friction of elastomers

In order to examine the adhesive component of sliding friction, Grosch (1963) examined the coefficient of friction of four rubbers sliding on plane glass, as a function of temperature and relative velocity. By using plane glass and a rubber slider Grosch was able to eliminate any mechanical interaction between the surfaces and to isolate the adhesive component. His results are shown in figure 6.9(a), and shifted along the log velocity axis to form a master curve at 20 °C in figure 6.9(b). In figure 6.9(b), the shifts (a_T) for the different temperatures, which were required to make the individual curves form one master curve, were chosen solely to give the best possible superposition. However, when the theoretical shifts required to bring the curves into superposition were calculated on the basis of the known T_g of the rubber and the WLF equation, it was found that the theoretical and experimental shifts agreed remarkably well, indicating that the frictional properties were governed by the same viscoelastic processes that have been seen in chapter 5 to control the rupture of rubber to substrate adhesive bonds.

Grosch also used the WLF transform to develop, from torsion pendulum measurements over a range of temperatures and frequencies, master curves relating the dependence of the loss modulus E'' and the loss factor tan ϕ to the frequency at the same reference temperature as the friction master curve. He then found that if V_S was the velocity at which maximum friction was developed and $f_{E''}$ was the frequency of the maximum in the loss modulus, then for all the rubbers studied

$$V_S/f_{E''} = 6 \times 10^{-7} \text{ cm} = \lambda.$$

The frequency of the maximum in the loss modulus curve corresponds to that of the major relaxation process and so it appears that, since λ represents the distance which one molecular segment could jump during the relaxation time, a possible explanation for the result is that molecular bonds are formed and then broken again when they have jumped as far as they can.

Activated rate process theory of dynamic rubber friction

Schallamach (1963) incorporated these ideas in a theory to explain the velocity dependence of rubber friction on the basis of Eyring's (Glasstone, Laidler & Eyring, 1941) activated rate process theory. Schallamach suggested that intermolecular bonds are formed between the surfaces and that as the surfaces move past each other these bonds cause molecular segments to jump in the direction of the relative motion. When the molecular segment can jump no further, the bond is broken and the segment then jumps back to its equilibrium position in the surface. The mean time

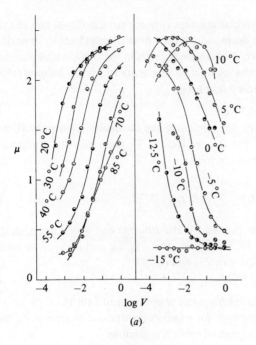

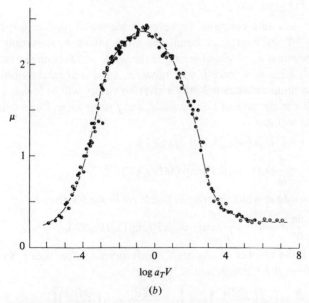

Fig. 6.9. (a) Variation of friction with speed for acrylonitrile-butadiene-rubber sliding over a hard surface. (b) Master curve for the friction data. (From Bowden & Tabor, 1966.)

during which a molecular segment is being stretched is $\bar{t}$, and the normal period of relaxation during which the segment jumps back to its equilibrium position is t_0 so that, if there are N_0 sites in the surface which are capable of forming bonds, the number of segments which are being stretched at any one time is given by N, where

$$N = N_0\ \bar{t}/(\bar{t} + t_0).\qquad(6.28)$$

If the force constant for stretching the molecular segment is M, and the relative velocity of the two surfaces is V, then the total frictional force between the surfaces is given by

$$F = \sum_{i=1}^{i=N} f_i = \sum_{i=1}^{i=N} MVt_i = NMV\bar{t},\qquad(6.29)$$

where f_i is the force exerted by the ith segment which has been stressed for a time t_i since its formation, and t is the mean life of an interfacial bond. Hence, combining (6.28) and (6.29)

$$F = N_0 MV\ \bar{t}^2/(\bar{t} + t_0),\qquad(6.30)$$

and it is necessary to calculate the dependence of $\bar{t}$ on V.

If the activation energy for a molecular segment to jump is ϵ_0, then the relaxation time for segmental motion is given by

$$(1/t_0) = k' \exp(-\epsilon_0/kT),$$

where k' is a rate constant. However, if segmental motion involves the breaking of an interfacial bond, then the effective activation energy barrier becomes $\epsilon_0 + w_a$ where w_a is the work of adhesion per molecule. The application of a force f_i will, however, assist segmental motion and if the mean jump distance is λ, the activation energy will be lowered by $f_i\lambda$. Hence, if r is the rate of bond scission and $\bar{f}$ is the mean force acting on a molecular segment

$$r = k' \exp\left[-(\epsilon_0 + w_a - \bar{f}\lambda)/kT\right]$$
$$= \frac{1}{t_0}\exp(-w_a/kT)\exp(\lambda MVt/kT),$$

hence the rate at which interfacial bonds are broken is given by

$$\frac{dn}{dt} = -nr = -\frac{n}{t_0}\exp(-w_a/kT)\exp(\lambda MVt/kT),\qquad(6.31)$$

where n is the number of original N bonds surviving after time t. So writing $u = \exp(-w_a/kT)$, integration of (6.31) yields

$$\frac{n}{n_0} = \exp\left(\frac{kTu}{t_0\lambda MV}\right)\exp\left[-\frac{kTu}{t_0\lambda MV}\exp\left(\frac{\lambda MVt}{kT}\right)\right],$$

where n_0 is the number of bonds formed initially. Since by definition

$$\bar{t} = \int\limits_0^\infty \frac{n}{n_0} \, dt,$$

$$\bar{t} = \frac{kT}{\lambda MV} \exp\left(\frac{kTu}{\lambda MVt_0}\right) \int\limits_{ukT/\lambda MVt_0}^\infty \frac{\exp(-q)}{q} \, dq, \qquad (6.32)$$

where $q = (kTu/\lambda MVt_0)\exp(\lambda MVt/kT)$. But $\int_x^\infty t^{-1}\exp(-t)dt$ is the exponential-integral function of $-x$, $\mathrm{Ei}(-x)$, and so (6.32) may be rewritten as

$$\frac{u\bar{t}}{t_0} = \frac{kTu}{\lambda MVt_0} \exp\left(\frac{kTu}{\lambda MVt_0}\right)\mathrm{Ei}\left(\frac{kTu}{\lambda MVt_0}\right).$$

Consequently, if (6.30) is written in the form

$$F(w_a) = \frac{N_0 kT}{\lambda} \frac{\lambda MVt_0}{kT} \left[\frac{(\bar{t}/t_0)^2}{(1 + \bar{t}/t_0)}\right]$$

$$= \frac{N_0 kT}{\lambda} \frac{\lambda MVt_0}{kT} \, \phi\left(\frac{kTu}{\lambda MVt_0}\right), \qquad (6.33)$$

the variation of F with V for various values of w_a can be represented by a plot of $\lambda F/N_0 kT$ against $\lambda MVt_0/kT$. Two curves for $w_a = 0$ and $w_a = 3kT$ are shown in figure 6.10.

It can be seen that the theory semi-quantitatively predicts the observed results and indicates that as the speed of sliding increases the force required to disrupt the bonds increases, but the number of effective bonds decreases

Fig. 6.10. Theoretical variation of frictional force (plotted as $\lambda F/N_0 kT$) with velocity (plotted as $\lambda MVt_0/kT$). (After Schallamach, 1963, p. 380.)

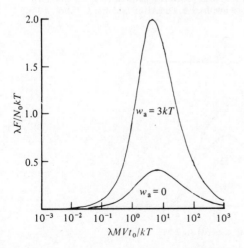

so that there is a peak in the frictional force at the velocity of sliding appropriate to the relaxation time for molecular segment movement.

The effect of the work of adhesion on the total frictional force can also be seen. The value of $w_a = 3kT$ was used because it had been found that such a value, equal to the energy of a van der Waals' bond, enabled Hatfield & Rathman (1956) to explain the time dependence of the strength of a polyisobutylene polymethyl methacrylate bond, and it can be seen that a change in w_a from 0 to $3kT$ raises the frictional force by a factor of about 5, but makes very little difference to the velocity at which maximum frictional force is developed.

Ludema & Tabor's (1966) explanation of the peak in the F versus log V curve has a similar basis but is couched in macroscopic terms. The contact area, Ω, is an inverse function of the modulus of elasticity and consequently it decreases sharply at velocities of sliding higher than that appropriate to segmental relaxations. The shear strength τ_s increases with strain rate but levels off at high strain rates (Smith, 1958); hence, combining the two results and shifting the variation of shear strength with velocity to higher strain rates, because of the strain rate magnification effect caused by concentrating all the strain in a surface layer approximately 10 nm thick, gives the result that the frictional force

$$F = \Omega \tau_s$$

goes through a maximum, as shown in figure 6.11.

Fig. 6.11. Curves showing variation of τ_s and Ω as functions of sliding speed assuming that the shear processes involved in τ_s are about 10^5 times as rapid as those involved in Ω. (After Ludema & Tabor, 1966, p. 336.)

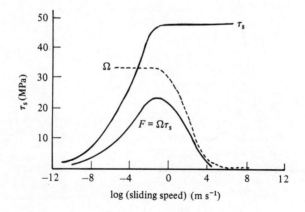

Surface energies and sliding friction of elastomers

Figure 6.10 shows the strong dependence of the sliding friction of a rubber on the work of adhesion between the rubber and the substrate. Savkoor (1974) has investigated the relationship linking the sliding friction between a rubber and a series of different indenters to the contact angle made by a liquid on the solid indenters.

Using the subscripts e for the elastomer, s for the indenting solid and l for the test liquid, the work of adhesion (equation (2.5b)) may be written as

$$(w_a)_{es} = 2\phi_1 (\gamma_e \gamma_s)^{1/2}, \tag{6.34}$$

where ϕ_1 is assumed to be a constant for all the solid–elastomer interactions. For the spreading of a liquid on the solid surface, Young's equation (2.17) gives

$$\gamma_s = \gamma_l \cos \theta + \gamma_{sl}, \tag{6.35}$$

and substituting for γ_{sl} from (2.5a)

$$\gamma_{sl} = \gamma_s + \gamma_l - 2\phi_2(\gamma_s \gamma_l)^{1/2}, \tag{6.36}$$

and so similarly assuming that ϕ_2 is a constant for all solid–liquid interactions the combination of (6.35) and (6.36) gives

$$\gamma_s^{1/2} = \gamma_l (1 + \cos\theta)/2\phi_2\gamma_l^{1/2}$$

or, substituting back in (6.34),

$$(w_a)_{es} = (\phi_1/\phi_2) \gamma_e^{1/2} \gamma_l^{1/2} (1 + \cos\theta),$$

so that there should be a simple relationship between the frictional force and $(\cos \theta)$. This relationship is shown in figure 6.12 (Savkoor, 1974) for wetting by two different liquids.

The sliding friction of plastics

For the majority of rigid plastics (and in this context the term 'rigid plastics' includes the three classes of materials, tightly cross-linked thermosetting resins, glassy polymers below their glass transition temperature and semi-crystalline polymers above the glass transition temperature) the coefficients of static and sliding friction are similar, i.e. the force required to initiate sliding is close to that which is required to maintain the relative velocity. Polytetrafluoroethylene homopolymer and linear polyethylene do, however, demonstrate a much lower value for the coefficient of sliding friction than for the coefficient of static friction and, because of this, these two polymers will be excluded from the general discussion which follows and considered separately in a later section.

McLaren & Tabor (1963) investigated the frictional properties of a wide range of polymers sliding on themselves over a range of relative velocities from 2 mm s^{-1} up to nearly 10 m s^{-1}. In general, the range over which the coefficient of friction varies with the relative velocity of the two surfaces is considerably less than that which is observed for the elastomers described previously. However, there is a general correlation between molecular mobility and frictional properties in that highly cross-linked thermosetting resins which have a restricted molecular mobility and which demonstrate very little strain rate sensitivity of their mechanical properties have a low coefficient of friction which varies little with the relative velocity of the surfaces. Amorphous thermoplastics below their glass transition temperature demonstrate both a rather higher coefficient of friction and a more marked velocity dependence. The velocity dependence is most marked for the semi-crystalline polymers, as can be seen in figure 6.13.

The frictional properties of any two surfaces are governed by the relative strengths of the interface and the materials on either side of the interface. Thus, if the interface is weaker than the material on either side, then on application of sufficient force, interfacial bonds are broken and the surfaces after separation are seen to be 'clean'. On the other hand, if the shear strength of one of the materials which makes up the friction pair is less than that of the interface, material close to the interface will be sheared, and after separation of the surfaces it can be seen that material transfer has taken place from one surface to the other. Thus Pooley & Tabor (1972) examined the friction between glass and a number of polymers. The frictional contacts were in the form of a hemispherically ended rod which was made to slide over a glass plate. Relative velocities between 20 μm s^{-1}

Fig. 6.12. Relation of contact angles to peak friction. *P* glass, PMMA, stone (granite), clean stainless-steel, *Q* polystyrene (steel, glass not cleaned), *R* polyethylene, *S* PTFE, I liquid drop water (H$_2$O), II liquid drop mercury (Hg). (From Savkoor, 1974, p. 99.)

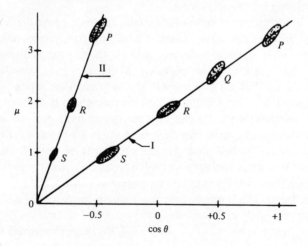

and 2 mm s^{-1} were used at temperatures from ambient to 150 °C. They found that for the amorphous polymers polystyrene and polymethyl methacrylate, and for the semi-crystalline polymers polypropylene and polychlorotrifluoroethylene at ambient temperatures there was no polymer transfer, but that at higher temperatures polymer transfer took place leaving a trail of lumpy fragments as much as 100 nm thick. The actual variation of the coefficient of friction on moving from the interfacial to the bulk shear regime depends upon the individual polymer. Although the shear strength of the interface decreases, because of the decrease in the modulus of the polymer the true area of contact increases and so the coefficient of friction may rise or fall depending upon the balance between these factors. In table 6.3 the diameter of the circle of contact is reported so that the shear stress which sustains sliding can be calculated, and it can be seen that whilst this falls with temperature the coefficient of friction may rise or fall.

When Pooley & Tabor compared the shear strength of the interface with the bulk shear strength of the polymers, they found that the ratio $(\tau_s)_{bulk}/(\tau_s)_{surf}$ varied between 2.6 and 3.9, and ascribed this to the possibility either that the sliding interface contained imperfections which were not present in the bulk specimens, or that the true area was considerably less than that which was calculated from the track width.

Pressure dependence of friction

If the factor controlling the frictional force is the shear strength of the polymer, then the pressure dependence of the yield point of the polymer should be reflected in its frictional properties. There is a considerable amount of evidence (Sternstein & Ongchin, 1969) that the

Fig. 6.13. Variation of friction with speed at room temperature for a number of polymers. (From McLaren & Tabor, 1968, p. 857.)

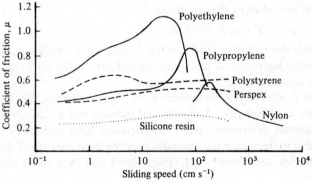

Table 6.3. *Variation of frictional shear stress with temperature*

Material	Temperature (°C)	μ	Diameter of track (mm)	$\tau_s = 4\mu X/\pi d^2$	Polymer transfer
Polystyrene	ambient	0.30	0.45	0.95	No
$X = 2.4$ N	50	0.35	0.50	0.90	No
	100	0.75	2.30	0.09	Yes
	150	0.45	4.50	0.01	Yes
Polymethyl	ambient	0.30	0.5	0.76	No
methacrylate	50	0.32	0.5	0.76	No
$X = 2.4$ N	100	0.20	1.2	0.09	Yes
	150	0.40	2.4	0.04	Yes
Polyvinyl	ambient	0.30	0.5	0.76	No
chloride	50	0.38	0.6	0.70	No
$X = 2.4$ N	100	0.30	1.2	0.13	Yes
	150	0.20	2.0	0.03	Yes
Polypropylene	20	0.27	0.48	1.5	No
$X = 4.9$ N	50	0.33	0.53	1.5	No
	100	0.34	0.60	1.2	No
Polychloro-	ambient	0.28	0.50	1.4	No
trifluoro-	50	0.40	0.60	1.4	No
ethylene	100	0.50	0.71	1.25	No
$X = 4.9$ N	150	0.36	1.0	0.45	Yes

appropriate yield criterion for a rigid polymer is one which takes the form of a von Mises criterion modified to depend upon the hydrostatic tension component of the stress. Towle (1974) has suggested, following Bridgeman (1935), that the shear strength of a polymer may be written in the form

$$\tau_s(P) = \tau_0 + \alpha P + \beta P^2 + \dots .$$

He has devised an experimental arrangement in which he can apply very high pressures and in which at high pressures the true area of contact is a constant. In such an arrangement the coefficient of friction varies with pressure according to the expression

$$\mu = \tau_s/P = \tau_0/P + \alpha + \beta P,$$

and since the τ_0 and α terms are the dominating terms, μ decreases from an initially high value at low pressures and then levels out to an approximately constant value.

Aird & Cherry (1978) examined the variation of the frictional force between a polyethylene hemispherical indenter and a glass plate, and by simultaneous determinations of the frictional force and area of contact were able to determine the dependence of the shear strength of the friction junction on the normal load. They found that they could explain the variation of the coefficient of friction with load if they assumed a pressure-dependent von Mises yield criterion, but that the value of the shear strength of the polymer at the interface was much lower than in the bulk. This effect could also be seen in the results of Pooley & Tabor (1972) and of Briscoe & Tabor (1975).

Surface energies and the sliding friction of plastics

Since the deformations involved in the sliding friction of plastics are so large, it is unlikely that it would be possible to obtain a quantitative relationship between the coefficient of friction and the work of adhesion in the same way that it was found for elastomers. However, equation (6.34) can be written using the subscripts 'p' for plastic substrate and 's' for solid rubber as

$$(w_a)_{ps} = 2\phi(\gamma_s \gamma_p)^{1/2},$$

so that for a constant slider

$$(w_a)_{ps} = (2\phi\gamma_s^{1/2})\gamma_p^{1/2}.$$

Lee (1974) examined the data of Tanaka (1961) for a series of polymers which, because they had very similar mechanical properties, might be expected to show a dependence of frictional characteristics on the surface

Fig. 6.14. Correlation between coefficient of static friction and critical surface tension for wetting. PMMA: polymethyl methacrylate, HDPE: high-density polyethylene, PVC-30: polyvinyl chloride-30 p.h.r. plasticiser, PVC-50: polyvinyl chloride-50 p.h.r. plasticiser.

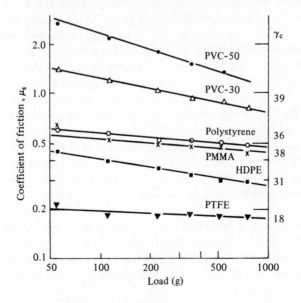

energy. Equation (2.25) suggests that in the absence of polar interactions γ_p may be approximated by γ_c, the critical surface tension for wetting, so Lee looked for a correlation between the coefficient of friction and the critical surface tension; the results shown in figure 6.14 suggest that such a correlation does exist.

Polytetrafluoroethylene and linear polyethylene

Polytetrafluoroethylene (PTFE) has long had the reputation for having exceptional properties as far as frictional properties are concerned, and this was for a long time ascribed to weak interfacial bonding between the polymer and the substrate. However, Makinson & Tabor (1964) showed that polymer transfer takes place and hence that the interfacial inter-actions may be strong even though the frictional forces are weak. Pooley & Tabor (1972) showed that when a fresh PTFE surface is made to slide over a glass substrate then the initial coefficient of friction is high ($\approx$0.1) and a lumpy transfer of polymer to the glass surface takes place in a very similar fashion to other polymers for which the interfacial shear strength is greater than the bulk shear strength. However, as the sliding process continues, the coefficient of friction falls to a much lower value (0.02-0.06) and the polymer transfer instead of being lumpy becomes very tenuous and of the order of molecular dimensions in thickness. Repeated traverses over the same track fail to reproduce the initial high static coefficient of friction, and neither does a traverse of fresh glass, provided that the direction of sliding over the polymer is in exactly the same direction as the first slide. If, however, the polymer is made to slide at 90° to the original direction, then the initial high coefficient of friction and lumpy transfer is observed. The explanation offered by Pooley & Tabor is that the initial sliding (with a high coefficient of friction) orients the polymer surface so that the poly-mer chains lie parallel to the direction of relative motion. PTFE, however, has a very smooth molecular profile so that the force required to draw a very thin film from the polymer is very low, and consequently the frictional force falls to a very low value. Repeated traverses on the same or on a fresh surface parallel to the original direction of motion develop only the frictional forces necessary to draw out the thin film. However, rotating the polymer through 90° means that it is no longer in a favourable orientation for drawing, and so the high initial coefficient of friction is again observed.

Linear polyethylene behaves in a very similar fashion, and in both these polymers the low friction regime stems from the ease with which a smooth profiled polymer molecule can be drawn from the surface. Paradoxically, the low friction of PTFE and linear polyethylene depends upon its high adhesion.

7   Deformation and friction

7.1   The mechanical interactions of surfaces
Friction caused by surface deformation

The frictional effects discussed in chapter 6 must all be associated with surface deformations since, even if the applied force normal to the interface is reduced to zero, the action of frictional forces parallel to the interface must bring about deformation parallel to the interface. However, even if adhesive interactions are reduced to negligible proportions, friction can still be experienced between two surfaces due to the mechanical inter-action of the surfaces, either by the juxtaposition of asperities on the two surfaces or by one surface indenting the other. The adhesive component of friction can be eliminated by lubrication, in which a liquid of low shearing stress remains interposed between the two surfaces, or it can be eliminated by the rolling of one surface over the other if both the formation and the disruption of the interface are thermodynamically reversible. The term 'deformation component of friction' has come to be applied to frictional effects associated with mechanical interactions of the surfaces and is used to differentiate these effects from those associated with the adhesive interaction of the surfaces, whether the disruption of the interface is thermodynamically reversible or not.

In the discussion of the adhesive mechanism of sliding friction in section 6.4, it was suggested that the deformations which were brought about by adhesive mechanisms of friction were of the order of 1 nm. In the discussion which follows it may be seen that the deformations which are associated with what is usually termed the deformation component of friction exceed 1 μm. Consequently, an alternative distinction between adhesive and deformation mechanisms of friction might be made in terms of the magnitudes of the deformations involved, and a dividing line set at 1 μm. However, it must be realised that there is in fact a continuous range of such deformations and the distinction between the two mechanisms is arbitrary but convenient.

The deformation component of friction arises because in order to bring about sufficient deformation of one or both surfaces to let them slide past each other a certain amount of energy must be expended. If the work which is done when two surfaces slide at constant relative velocity a distance Δ relative to each other is given by W_f, then the measured frictional force must be given by F, where

$$F\Delta = W_f. \tag{7.1}$$

In general, the work which is expended in overcoming frictional forces may be expended either in bringing about plastic or viscoelastic deformation of the surfaces, and this forms the basis of a distinction between two different forms of the deformation component of friction. If the energy is dissipated in bringing about permanent plastic deformation of one or both surfaces as the harder material ploughs through the softer then this is termed 'ploughing friction'. If the mechanical interactions cause viscoelastic deformations and energy is dissipated by molecular relaxation processes then this is termed 'hysteresis friction'. The term 'hysteresis friction' is of course as much a misnomer as the term 'deformation friction', since all deformations of viscoelastic solids dissipate energy by a hysteresis mechanism (Moore & Geyer, 1972), but the term will be used for just those cases where energy is dissipated by macroscopic viscoelastic deformations of the polymer. These deformations will extend over a greater volume of the polymer than those involved with the adhesion mechanism of friction.

The nature of hysteresis friction

A simple demonstration of the hysteresial component of friction is afforded by the example shown in figure 7.1(a), in which a hard sphere moves through the surface of an elastomeric solid. Because the material in front of the indenter has to be compressed, work has to be expended to overcome the materials resistance to compression. Behind the indenter, viscoelastic recovery will assist the forward motion of the indenter, but because of the viscoelastic losses associated with the material being subjected to a strain cycle, some of the energy input will be dissipated and the amount of work 'recovered' from the system can never equal the input. The difference between the input work and the work recovered during the relaxation phase represents the work expended in overcoming the hysteresial component of friction.

If, in moving a distance Δ, the work done in compressing the material ahead of the indenter is W_i and a fraction α of this is dissipated, then in equation (7.1)

Fig. 7.1. Deformation friction. (a) Hysteresis friction, (b) Ploughing friction.

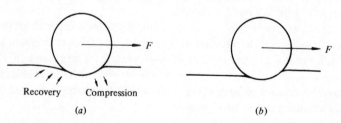

Recovery   Compression

(a)   (b)

$$F\Delta = \alpha\,W_i$$

or

$$F = \alpha\,W_i/\Delta. \tag{7.2}$$

Since α is a function both of the relative values of the frequency of the stress cycle to which the polymer is subjected and of the frequency for maximum losses in the polymer, α will vary with the relative velocity of the two surfaces.

It was pointed out previously that the adhesive component of friction could be reduced to negligible proportions by efficient lubrication: consequently, since the deformation of a polymer surface which may be brought about by the lubricated sliding of a rigid indenter over it is precisely similar to the deformation brought about by a spherical indenter of the same dimensions rolling over it, the observed friction should be the same whether rolling or sliding has occurred. That this is in fact the case was reported by Greenwood & Tabor (1958), who measured the coefficient of friction for steel spheres both sliding and rolling over a styrene-butadiene-rubber (SBR) co-polymer which had been lubricated by means of a soap film. The results are shown in figure 7.2 and can be seen to be coincident for the two cases.

The nature of ploughing friction

The simplest representation of ploughing friction can be seen in figure 7.1(b) in which, when a hard sphere moves through a plastic surface, a groove is left behind. The frictional force in such a case is simply the product of the cross-sectional area of the groove and the yield pressure of the plastic. It should, however, be noted that although the ploughing term which may arise when, for example, a pointed indenter is first dragged through a soft surface may be large, the frictional force will rapidly fall to

Fig. 7.2. The similarity between results for rolling and sliding friction. (After Bowden & Tabor, 1964, p. 262.)

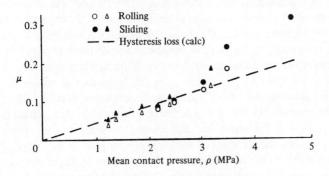

Mean contact pressure, p (MPa)

lower values. This is because if the indenter makes repeated traverses over the same track then orientation of the molecules in the surface of the track will lead to work hardening of the track and the smaller indentations will give rise to a lower value for the frictional force. Alternatively, if the indenter slides over an extended surface the indenter itself will soon get worn down to a smoother profile which causes less damage in the counter surface (Bowden & Tabor, 1966).

7.2 Hysteresis friction
Occurrence of the deformation mechanisms of friction

For many materials the deformation component is negligible in comparison with the adhesive component of friction. However, for elastomers in the range where they have high hysteresis losses, and particularly under lubricated conditions, the hysteresial component may become dominant.

The effect of the relative velocity of the two surfaces on the magnitude of the coefficient of friction was discussed in section 6.4, and the results for the sliding friction of an acrylonitrile-butadiene-rubber on a smooth surface (Grosch, 1963) were shown in figure 6.9. In figure 7.3 these results are compared with the results obtained when the rubber slides over clean silicon carbide paper (180 grit size, so that the average spacing between asperities in 0.14 mm) and when the rubber slides over similar silicon carbide paper which has been dusted with magnesium oxide powder. It can be seen that when the rubber slides over a silicon carbide paper surface, then a second peak occurs at higher velocities and this second peak can be attributed to the deformation mechanism of friction. When the silicon carbide paper is dusted with magnesium oxide powder, this acts as a lubricant and inhibits the formation of interfacial bonds, thus suppressing the peak in the coefficient of friction versus log velocity curve which was attributed in chapter 6 to the adhesion component of friction.

In the case of adhesive friction it was noted that for each rubber the ratio $V_S/f_{E''}$ was a constant, equal to 6 nm, and this distance was assumed to be the distance of an intermolecular jump. In the case of the deformation component of friction a similar effect was observed. If V_R was the relative velocity corresponding to the peak in the coefficient of friction versus log velocity curve which could be ascribed to the deformation mechanism of friction, then the ratio V_R/f_t, where f_t is the frequency at which the dynamic loss factor $\tan\phi$ $(= E''/E')$ achieves its maximum value, is also a constant. The constant in this case was found to equal 0.15 mm, which is sufficiently close to the value for the spacing between the asperities (0.14 mm) to justify the ascribing of the high-velocity peak in the friction-velocity relationship to the deformation of the elastomer by the movement of the asperities on the counter surface through the elastomer surface.

Quantitative treatment of hysteresis friction

Two general methods have been utilised in order to attempt the calculation of the theoretical value of the hysteresial component of friction. Greenwood, Minshall & Tabor (1961) have calculated the work done when a counter surface of given shape compresses the elastomer, assuming a perfectly elastic deformation. They have then allowed for the hysteresis effects by assuming that a certain fraction of this energy is dissipated viscoelastically. Flom & Bueche (1959) assumed a very specific constitutive equation for the elastomer and calculated the stress distribution under a moving counter surface and hence the force resisting forward motion of the indenter. Both methods of calculation give results which have the same form, but they highlight different aspects of the frictional process and so both will be detailed below.

The form of the relationship between the frictional force and (say) the applied load is a function of the shape of the counter surface, so in order to simplify the calculations shown below, the friction of a rigid cylinder rolling or undergoing well lubricated sliding over a rubber surface will be examined so that in plane strain only two dimensions need to be considered.

Fig. 7.3. Master curves for the coefficient of friction of an acrylonitrile-butadiene-rubber on four surfaces, plotted against reduced velocity (velocity shifted to 20 °C): --- wavy glass, ----- polished stainless-steel, — clean silicon carbide, ······ dusted silicon carbide. (From Grosch, 1963, p. 29.)

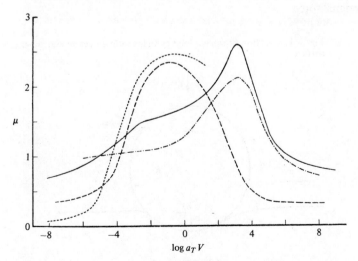

Calculation of the hysteresis friction from the energy input

Figure 7.4 shows a section through a long cylinder rolling over an elastomeric surface. The analysis by Hertz (Timoshenko & Goodier, 1970) gives the distribution of pressure under the cylinder as

$$P = \frac{2W}{\pi a} \left(1 - \frac{x^2}{a^2}\right)^{1/2}, \tag{7.3}$$

where W is the applied load per unit length and the meaning of the other symbols is shown in figure 7.4.

The displacement of the surface at any point $|x| < a$ is given by

$$z = z_0 - x^2/2R,$$

where z_0 is the displacement of that portion of the rubber surface immediately below the centre of the cylinder. If the cylinder moves forward a distance Δ then the change in displacement at x is given by $\Delta(\partial z/\partial x) = \Delta x/R$. Hence the work done by the pressure over a strip of width dx will be $-Pdx(\Delta x/R)$ and the total work done by the front half of the cylinder is given by

$$W_i = \frac{\Delta}{R} \int_0^a Pxdx = \frac{2}{3} \frac{\Delta Wa}{\pi R}. \tag{7.4}$$

The analysis by Hertz also yields

$$a = \frac{2}{\pi^{1/2}} \left[\frac{WR(1 - \nu^2)}{E}\right]^{1/2}. \tag{7.5}$$

So that if a fraction α of the input energy is dissipated viscoelastically, substituting equations (7.4) and (7.5) in (7.2) yields for the rolling frictional force

Fig. 7.4. A section through a long cylinder rolling over an elastomeric surface.

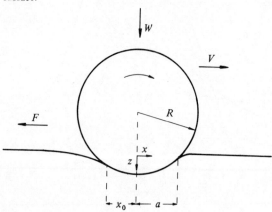

$$F_r = \frac{4}{3} \frac{W^{3/2}}{\pi^{3/2}R^{1/2}} \left(\frac{1-\nu^2}{E}\right)^{1/2} \alpha, \tag{7.6}$$

and hence the coefficient of friction of the cylinder, μ_r^c, is given by

$$\mu_r^c = F_r/W = \frac{4}{3} \alpha \left(\frac{1-\nu^2}{\pi^3 RE}\right)^{1/2} W^{1/2}. \tag{7.7}$$

By a similar calculation it may be shown that for a rigid sphere rolling over a plane, if μ_r^s is the coefficient of friction

$$\mu_r^s = \frac{3}{16} \left(\frac{3}{4}\right)^{1/3} \alpha \left(\frac{1-\nu^2}{R^2 E}\right)^{1/3} W^{1/3}, \tag{7.8}$$

from which it can be seen that for both rolling cylinders and rolling spheres Amontons' laws are not obeyed in that the coefficient of friction increases with applied load. The predictions of equations (7.7) and (7.8) were shown to be correct by the results of Greenwood *et al.* (1961) shown in figure 7.2.

In equations (7.7) and (7.8) the constant α represents the fraction of the input energy which is dissipated by viscoelastic relaxations; α will of course vary with the speed of rolling since this will alter the effective frequency of the strain cycle to which the rubber is subjected. Greenwood *et al.* attempted to estimate from mechanical measurements what α should be. They demonstrated the very important point that the energy dissipated by hysteresis is not determined primarily by the total amount of energy supplied to the specimen, but rather by the nature of the strain cycle to which the rubber is subjected. They showed, for example, that as the sphere rolled over an element of the rubber, this volume element was subjected to deformations similar to those indicated approximately in figure 7.5. It was found that although the stored elastic energy at B was very similar to that at C, because of the shearing which took place between B and C there was

Fig. 7.5. Deformation of elements at a depth a below the surface in the rolling of a cylinder over a flat surface. (From Bowden & Tabor, 1966, p. 306.)

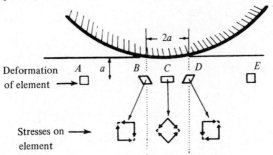

a considerable dissipation of energy by hysteresis mechanisms during the deformation from B to C. Hence it appears that the input of energy during the initial deformation is no guide to the total hysteresis losses, which must be calculated for each individual stress cycle. However, Greenwood *et al.* were able to determine the hysteresis losses observed when they subjected thin-walled tubes to tension cycles, torsion cycles and combinations of tension and torsion and so predict with some accuracy the observed frictional forces due to hysteresis losses.

A further point which arises from the analysis of Greenwood *et al.* is that the full losses only occur when a piece of rubber is taken through a complete strain cycle. This implies that the losses experienced in the early stages of rolling and hence the frictional force at the commencement of rolling are much less than when the rolling motion is fully developed. The increase in rolling friction as a function of distance rolled is shown in figure 7.6, and this may do much to explain the ease with which wheel spin occurs when impatient drivers try to move away too quickly when traffic lights change.

Calculation of hysteresis friction from the stress cycle

The method adopted by Flom & Bueche (1959) for the calculation of the frictional force when a sphere travels across an elastomeric surface either by rolling or by well lubricated sliding, was to consider the forces exerted by the various portions of the rubber on the sphere assuming that the rubber could be regarded as a simple array of Voigt elements. Since in exaggerated form the profile of the rubber will be similar to that shown in figure 7.7, the unbalanced force exerted over the front portion of the sphere will be responsible for the frictional opposition to movement.

Fig. 7.6. Increase of rolling friction F as a function of the distance u rolled.

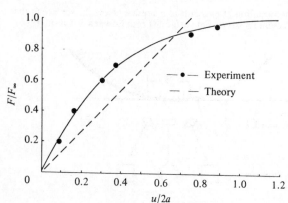

The profile proposed for the forward half of the indentation does of course differ markedly from the Hertzian profile discussed in chapter 6, but this is because the system is operating under non-equilibrium conditions when the rolling sphere is being squashed up against the rubber which is in front of it.

As in the case of the method adopted by Greenwood *et al.*, the calculations of the hysteresial component of friction by the method of Flom & Bueche are considerably simpler for the two dimensional case of a long cylinder rolling over a surface, and this calculation has been reported by Moore & Geyer (1972) in an extensive survey of this field.

If at any point P on the interface between a rolling cylinder and an elastomer the pressure is p, then if this pressure is supposed to be supported by a Voigt element

$$p = k(Gz + \eta \dot{z}), \tag{7.9}$$

where z is the vertical displacement of the surface of the elastomer from its undisturbed position, G and η are the constants describing the Hookean and Newtonian components of the Voigt element and k is a constant of proportionality which has dimension (length)$^{-1}$.

If x is the horizontal distance between P and the point vertically beneath the centre of the cylinder, then the displacement of P from its undisturbed position is given by

$$z = (R^2 - x^2)^{1/2} - (R^2 - a^2)^{1/2}.$$

Hence, since $\mathrm{d}z/\mathrm{d}t = -(\mathrm{d}z/\mathrm{d}x)(\mathrm{d}x/\mathrm{d}t) = v\,\mathrm{d}z/\mathrm{d}x$, where v is the forward velocity of the cylinder, $\mathrm{d}x/\mathrm{d}t$, then

$$\dot{z} = vx(R^2 - x^2)^{-1/2}.$$

Consequently, from equation (7.9)

$$p = k\{G[(R^2 - x^2)^{1/2} - (R^2 - a^2)^{1/2}] + \eta vx(R^2 - x^2)^{-1/2}\}. \tag{7.10}$$

Fig. 7.7. Coordinate system for a rolling cylinder.

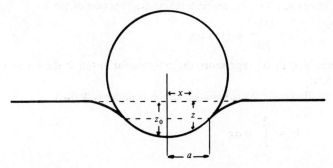

For light loads $a/R \ll 1$, $x/R \ll 1$, and so expanding equation (7.10) and neglecting terms higher than x^2/R^2 or a^2/R^2 yields

$$p = (kGR/2)\, [a^2/R^2 - x^2/R^2 + 2\eta vx/GR^2]. \tag{7.11}$$

Solving equation (7.11) for the condition $p = 0$ gives the value of $-x$ at which the rubber just separates from the rear edge of the rolling cylinder. If this is termed x_0,

$$x_0^2 - 2(\eta/G)\,vx_0 - a^2 = 0,$$

i.e.

$$x_0 = (\eta/G)v - [(\eta/G)^2 v^2 + a^2]^{1/2}, \tag{7.12}$$

from which it can be seen that the cylinder–elastomer contact is only symmetric when the cylinder is stationary.

For a Voigt element subjected to a cyclic strain of frequency ω, the phase factor ψ is given by $\tan \psi = \eta\omega/G$. For an element in the surface of the rubber ω may be approximated by $2\pi v/a$, and so it can be seen that $\tan \psi = 2\pi(\eta v/Ga)$, hence substituting in (7.12)

$$\frac{x_0}{a} = \frac{1}{2\pi}\,[\tan \psi - (1 + \tan^2 \psi)^{1/2}]$$

In order to calculate the frictional force due to hysteresis, F_r, moments may be taken around the centre of the cylinder yielding

$$F_r R = \int_{-x_0}^{a} px\,\mathrm{d}x,$$

i.e.

$$F_r = \frac{kG}{2R^2} \int_{-x_0}^{a} [a^2 - x^2 + (2\eta v/G)x]\,x\,\mathrm{d}x$$

$$= \frac{kG}{2R^2}\,a^3\left\{\left[1 - \frac{x_0}{a}\right] - \frac{1}{3}\left[1 - \left(\frac{x_0}{a}\right)^3\right] + \left(\frac{2\eta v}{Ga}\right)\left[1 - \left(\frac{x_0}{a}\right)^2\right]\right\}, \tag{7.13}$$

or, since each term in square brackets is a function of $\tan \psi$,

$$F_r = \frac{kGa^3}{2R^2}\,\phi_1(\tan \psi),$$

where $\phi_1(\tan \psi)$ represents the polynomial in $\tan \psi$ shown by equation (7.13).

Similarly, if the load applied to the cylinder is W then

$$W = \int_{-x_0}^{a} p\,\mathrm{d}x$$

$$= \frac{kG}{2R} \int_{-x_0}^{a} [a^2 - x^2 + (2\eta v/G)x] \, dx$$

$$= \frac{kGa^2}{2R} \phi_2(\tan \psi),$$

where $\phi_2(\tan \psi)$ is another polynomial in $\tan \psi$.

Hence the coefficient of friction for the hysteresial component is given by

$$\mu_r^c = \frac{F}{W} = \frac{a}{R} \phi_3 (\tan \psi), \tag{7.14}$$

where ϕ_3 is a function solely of $\tan \psi$. However, a is given by the Hertzian expression (equation (7.5))

$$a = \frac{2}{\pi^{1/2}} \left[\frac{WR(1 - v^2)}{E} \right]^{1/2}, \tag{7.5}$$

and so substituting in (7.14)

$$\mu_r^c = \frac{2}{\pi^{1/2}} \left[\frac{W(1 - v^2)}{RE} \right]^{1/2} \phi_3(\tan \psi),$$

which demonstrates the same dependence of μ_r on W, R and E as was derived by Greenwood *et al.* and reported in equation (7.7). If $\phi_3(\tan \psi)$ is evaluated for various speeds of rolling it can be shown that the variation of the hysteresial component of friction with velocity assumes the form shown in figure 7.8, which agrees with Grosch's experimental results, shown in figure 7.3.

If the similar calculations are carried out for the case of a sphere rolling over a plane elastomeric surface, then the results are found to be that

Fig. 7.8. Rolling friction of a sphere on an elastomeric material.

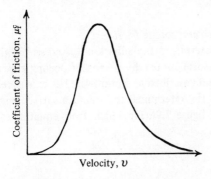

$$\mu_r^s = \frac{a}{R} \phi_4 (\tan \psi),$$

and substituting for a from equation (6.4) yields

$$\mu_r^s = \left[\frac{1 - \nu^2}{R^2 E} \right]^{1/3} W^{1/3} \phi_5 (\tan \psi),$$

which is again compatible with the treatment of Greenwood *et al.*, equation (7.8).

7.3 Adhesion hysteresis friction
Occurrence of adhesion hysteresis friction

Adhesion hysteresis friction is a cumbersome term, but is one which has to be introduced because in some circumstances effects due to the adhesive component of friction described in chapter 6 and effects due to the hysteresis component of friction described in the first two sections of this chapter do not act independently.

The experiments of Johnson, Kendall & Roberts (1971), described in section 6.2, were concerned with the formation of a glass–rubber interface under equilibrium conditions, so that the surface energy terms which were involved could be written as the components of the thermodynamic work of adhesion. Kendall, (1974) however, has extended this work on the formation of an interface between two solids to rates of formation and disruption which represent non-equilibrium conditions. Under these circumstances, the energy necessary to disrupt the interface may no longer be equal to the thermodynamic work of adhesion, but must be termed the fracture surface energy and analogously, since the energy released on formation of the interface may also no longer be the work of adhesion, it will be termed the manufacture surface energy. If the manufacture surface energy is less than the fracture surface energy then this 'hysteresis of adhesion' can give rise to large energy losses as a solid moves over a rubber surface and so give rise to a frictional mechanism, which will be termed adhesion hysteresis friction.

Adhesion hysteresis and rolling friction

Kendall (1975) determined the adhesion hysteresis for glass and a lightly cross-linked (i.e. soft) rubber by means of peeling experiments similar to those of Deryaguin described in section 4.3. His experimental set up is shown in figure 7.9. He determined the fracture surface energy R_b using the system shown in figure 7.9(*a*) in which, from equation (4.17), it can be seen that

$$R_b = \frac{W_b g}{w},$$

Fig. 7.9. Determination of the fracture surface energy and manufacture surface energy. (From Kendall, 1975, p. 354.)

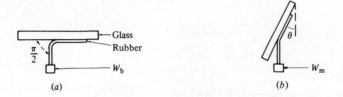

where w is the width of the rubber strip. Similarly he could determine R_m, the manufacture surface energy, using the experimental arrangement of figure 7.9(b) in which the glass plate was tilted until the rubber spontaneously pulled itself onto the glass surface under the influence of surface interactions. Again from equation (4.17), it can be seen that R_m is given by

$$R_m = \frac{W_m g}{w} (1 - \cos \theta).$$

By varying the experimental parameters, Kendall could vary the rate of crack propagation (either positive or negative) and hence could obtain the experimental results shown in figure 7.10, in which the variation of the adhesion hysteresis with the speed of the moving crack could be demonstrated.

Fig. 7.10. Manufacture surface energy, R_m, and fracture surface energy, R_b, as a function of crack speed (always considered positive). (From Kendall, 1975, p. 355.)

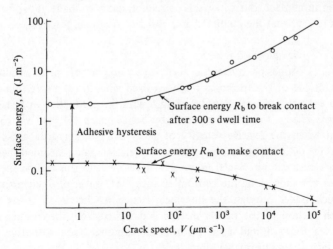

Fig. 7.11. Comparison of steady rolling friction results with predictions of the adhesion hysteresis theory. (After Kendall, 1975, p. 356.)

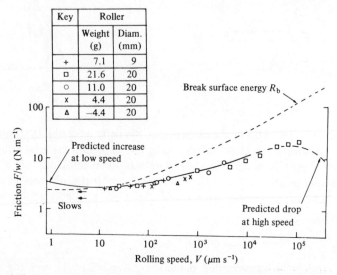

Key	Roller	
	Weight (g)	Diam. (mm)
+	7.1	9
□	21.6	20
○	11.0	20
x	4.4	20
Δ	−4.4	20

Kendall then measured the frictional force for a glass cylinder rolling over a plane rubber surface of the same composition as had been used previously, as a function of the rolling velocity. If the frictional force arises solely from adhesion hysteresis effects then, from equation (7.1),

$$F\Delta = w(R_b - R_m)\Delta,$$

where R_b and R_m can be determined from figure 7.10. In fact R_b has to be corrected for the 'dwell time', i.e. the time that the glass and rubber have been in contact, but when this correction has been made, it can be seen in figure 7.11 that the predicted and observed frictional forces agree well.

Schallamach waves

Adhesion hysteresis also plays a major role in determining the magnitude of the frictional force involved when (say) a soft rubber slider moves over a smooth counter surface with the production of Schallamach waves. These were first reported by Schallamach (1971), who observed that when the relative velocity of slider and counter surface exceeded a certain value which was specific to the friction pair concerned then a new mechanism of interfacial movement developed. In this regime the rubber does not slide along the counter surface, but 'waves of detachment' form which travel through the interface. Hence, as the slider moves forward each section of the rubber in turn is detached from the counter surface, moves forward and then reforms an adhesive bond with the counter surface at a point further along it.

The mechanism of the formation of Schallamach waves has been well described by Barquins & Courtel (1975) for the case of a rigid indenter travelling along the surface of a soft rubber. The model which they propose is shown in figure 7.12, in which buckling occurs at the leading edge of the slider and the fold increases in size until it cannot be accommodated, and so a wave of detachment forms which moves backwards along the indenter as the indenter moves forward. The relationship between the number of such waves and the relative velocity of the two surfaces has been calculated by Barquins, Courtel & Thirion (1974). They suggested that if the width of the wave as measured on the glass slider surface was l, then the length of the curved rubber surface would be S, which would be equivalent to δS of uncompressed rubber surface. Consequently if the slider velocity is V, the wave velocity is v and there are N waves traversing a total length of contact L, it can be seen that

$$VL = Nv\,\delta S.$$

If n is the number of waves generated per second, then $n = NV/L$ and so

$$V^2 = nv\delta S,$$

and typical values for n and v are shown in table 7.1 (Schallamach, 1971).

Briggs & Briscoe (1975) have shown that in the regime where Schallamach waves are generated, the adhesion hysteresis associated with the disruption of the interface is sufficient to account for all the frictional

Fig. 7.12. Schematic diagram of the evolution of the profile of a rubber surface and the formation of a Schallamach wave. (After Barquins & Courtel, 1975, p. 143.)

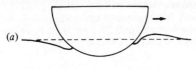

(a)

(b)

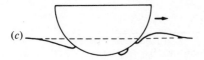

(c)

Table 7.1. *Hard slider on polyisoprene rubber*

Sliding speed (mm s^{-1})	Waves generated per second	Wave velocity (mm s^{-1})
0.24	5.6	8.6
0.43	11.7	10.7
0.93	23.6	14.7

forces observed. They examined the development of Schallamach waves when a rubber slider travelled over a plane glass surface, and then determined the fracture surface energy when the same rubber slider was pressed against a piece of glass and then pulled away from it at different rates. Briggs & Briscoe suggested that each time a wave of detachment travelled through the interface, an amount of energy $A(R_b - R_m)$ must be dissipated if A is the area of contact of slider and counter surface. Consequently, if n waves are generated per second

$$FV = A(R_b - R_m)n,$$

where F is the frictional force and V the relative velocity of the two surfaces.

Consequently the adhesion hysteresis, calculated as an energy dissipation per unit area, may be calculated from the direct fracture and manufacture surface energy determinations and from the observation of Schallamach waves. The results from both methods of calculation are shown in figure 7.13, and the good agreement between the two methods indicates the validity of the authors' hypothesis.

Fig. 7.13. Adhesion hysteresis for Dow Corning dielectric gel (15% curing agent): ○ measured directly (the continuous line refers to these points), ▲ deduced from Schallamach waves for 0.1 N normal load and two different sliding speeds. (From Briggs & Briscoe, 1975, p. 361.)

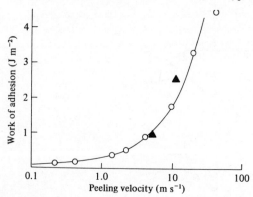

7.4 Ploughing friction and wear

The occurrence of ploughing friction

Ploughing friction occurs when a relatively hard asperity on one surface indents a softer counter surface causing general yielding of the softer surface, and then when sliding commences the hard asperity brings about further yielding in the counter surface parallel to the direction of motion. The ploughing component of friction may be relatively easily calculated, and in order to compare the results for ploughing friction with those for hysteresis friction the initial calculation for the ploughing component of friction will be carried out for a rigid spherical asperity moving through the surface of a softer polymer, whose hardness is given by h, as in figure 7.14 (Bowden & Tabor, 1966).

Since in the ploughing friction regime the counter surface is assumed not to recover after having been traversed by the indenter, the whole of the load must be supported by the material of the counter surface which is under the front surface of the indenter. Hence, if h is the effective hardness of the counter surface

$$\frac{1}{2} \pi a^2 = \frac{W}{h}. \tag{7.15}$$

The cross-sectional area of the groove cut by the spherical indenter, figure 7.14(a) is given by $A'' = \frac{1}{2}R^2(2\theta - \sin 2\theta) \approx a^3/16R$. Now the frictional force must in this case be caused by the yielding of the material in front of the indenter, and so

$$F = A''h \approx h\, a^3/16R,$$

so that substituting from (7.15)

$$F = (32h\pi^3R^2)^{-1/2} W^{3/2},$$

i.e.

$$\mu = k' W^{1/2}, \tag{7.16}$$

so that it may be seen that for the ploughing component of friction Amontons' laws are not obeyed and the dependence of the coefficient of friction on the applied load is stronger than that of any other component of friction.

Fig. 7.14. Ploughing friction. (a) Spherical indenter, (b) Conical indenter.

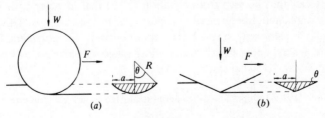

Abrasive wear

If in the ploughing friction regime all the material displaced by the indenter was to be removed from the softer surface as wear fragments, it would be comparatively simple to calculate the rate of wear of the surface (Lancaster, 1973). If a hard conical asperity indents a softer surface, as in figure 7.14(b), then the area supporting the load is $\pi a^2/2$ and the cross-sectional area of the track is given by $a^2 \tan \theta$. Consequently, by analogy with the derivation of equation (7.16), the coefficient of friction is given by

$$\mu = \frac{F}{W} = \frac{2ha^2 \tan \theta}{\pi a^2 h} = \frac{2}{\pi} \tan \theta. \tag{7.17}$$

By the same reasoning, if V is the volume of material removed when the indenter moves a distance L,

$$V = La^2 \tan \theta = \frac{2LW \tan \theta}{\pi h},$$

where h is the hardness. The 'volumetric wear rate', K_v,[†] which is the volume of material removed per unit distance of sliding, is then given by

$$K_v = \frac{2W \tan \theta}{\pi h}, \tag{7.18a}$$

or since the assumption that all the material displaced is removed is not justified, equation (7.18a) is better written

$$K_v = \frac{kW \tan \theta}{h}, \tag{7.18b}$$

where k is an arbitrary constant.

Two other definitions of wear rate are commonly used; the specific wear rate K_s is simply K_v/W; the energetic wear rate K_e is the volume of material used per unit work of friction; and so combining equations (7.17) and (7.18) it can be seen that if the simple ploughing mechanism proposed above applies to the situation being considered

$$K_e = \frac{2LW \tan \theta}{\pi h F L} = \frac{k'}{h}, \tag{7.19}$$

where k' is an arbitrary constant.

It can thus be seen from equation (7.19) that in general the energetic wear rate should be inversely proportional to the hardness. This relationship is usually only obeyed very approximately for polymers, but has a more general applicability over a wider range of materials, as can be seen from figure 7.15 (Lancaster, 1973).

[†]Moore (1972b) adopts a modified definition of the volumetric wear rate.

Figure 7.15 includes plastics as representatives of polymeric materials. Rubbers would, in fact, be likely to give a better performance because the deformations brought about during abrasive contact with a counter surface are more likely to be elastic.

When an elastomer is abraded, it commonly develops a pattern of ridges perpendicular to the direction of travel of the indenter (Schallamach, 1958) and the development of this pattern is associated with wear. The mechanism of wear in an elastomer is likely to be very different from that of a plastic, since plastic deformation can not take place in an elastomer. Removal of wear debris from an elastomer must be by a process of fracture of fragments away from the surface, and Champ, Southern & Thomas (1974) have proposed a semi-quantitative treatment for abrasive wear of rubber in terms of the fracture surface energy for cohesive tearing of the rubber.

Fig. 7.15. Variation of wear rate with hardness for abrasion on coarse (100) carborundum paper: 1 polyethylene (LD), 2 PTFE (polytetrafluoroethylene), 3 polypropylene, 4 acetal, 5 PVC, 7 FEP (tetrafluoroethylene-hexafluoropropylene co-polymer), 8 PMMA, 9 PPO (polypropylene oxide), 10 PTFCE (polytrifluorochloroethylene), 11 polycarbonate, 12 nylon 6, 13 nylon 6,6. (From Lancaster, 1973, p. 301.)

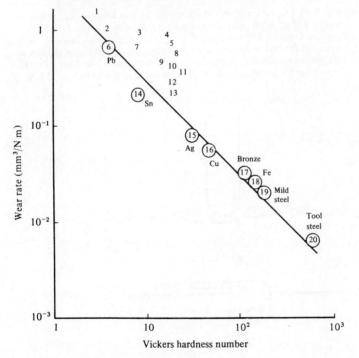

Champ *et al.* proposed that the deformation of an abrasion pattern which is schematically shown in figure 7.16(*a*) may be modelled as shown in figure 7.16(*b*). From the treatment of the 'trouser-leg' tear test shown in figure 4.7(*c*) and the development of equation (4.17) it can be seen that

$$R = F/b,$$

where *b* is the breadth of the abrading surface and *F* is the force applied to the projecting tongue of the abrasion pattern. Now the rate of cut growth in a fatigue situation has been examined by a number of authors (e.g. Thomas, 1958), and it has been found that for many materials the rate of cut growth (in mm/cycle) is given by *r*, where

$$r = BR^{\alpha},$$

and *B* and α are constants. Under equilibrium conditions, i.e. when the abrasion pattern has assumed its final form and is not changing in shape, a thickness *r* must be removed from the surface of the rubber each cycle in order that the abrasion cut will remain of constant depth. Thus the volumetric wear rate for such a system is given by $K_v = rb$, so

$$K_v = Bb^{\alpha - 1}F^{\alpha},$$

where α and *B* can be determined from cut growth experiments and *F* from friction experiments. Since α varies from about 2 for a natural rubber

Fig. 7.16. Derivation of an expression for pattern abrasion. (*a*) Schematic diagram of abrasion pattern and its deformation by a blade. Force *F* is applied to blade. Points such as *P* and *Q* are where crack growth is assumed to occur. *P* is the re-entrant angle in the undeformed abrasion pattern, and *Q* represents a similar element deformed by the blade. (*b*) Model for crack growth under abrading force *F*. (From Champ *et al.*, 1974, pp. 135–6.)

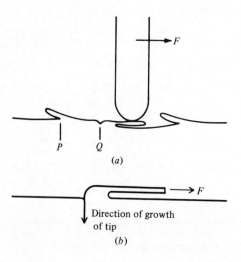

to 4 or more for non-crystallising rubbers such as SBR, it can be seen that the volumetric wear rate is strongly dependent upon the frictional force exerted between the rubber and the indenter.

Fatigue wear

If an indenter travels over a polymer surface, then successive portions of that surface are subjected to a strain cycle, and if the process is repeated often enough then fatigue failure will occur and portions of the polymer will be detached from the surface. This fatigue wear, as it is termed, is probably the mildest form of wear of all those discussed in this chapter, but because it occurs whenever two surfaces are brought into contact and continues as long as relative motion continues, it accounts for the majority of the wear of polymer surfaces.

A simplified treatment of the fatigue of a rubber surface has been developed by Reznikovskii (1960), and has been presented in an expanded form by Moore (1972b, p. 261). The rubber is assumed to travel over a rough surface which consists of a regular array of sinusoidal asperities, as shown in figure 7.17. If it is assumed that the mean thickness of a layer which is removed by fatigue wear is h and that on sliding a distance L over the surface, N such layers each of area A are removed, then the volume of rubber lost is given by

$$V = NAh,$$

and the work done by frictional forces is given by $L\mu W$, so that the energetic wear rate K_e is

$$K_e = \frac{NAh}{L\mu W}.$$

If λ is the wavelength of each asperity and it requires M cycles to bring about failure of the rubber, $NM\lambda = L$, and so

$$K_e = \frac{Ah}{\lambda M \mu W}. \tag{7.20}$$

Now h will vary as the depth of penetration δ of the asperities and as the roughness of the surface. From the Hertz equation (equation (6.2))

Fig. 7.17. Derivation of an expression for fatigue abrasion. (From Moore, 1972b, p. 261.)

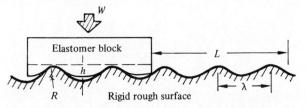

$$\delta \propto w^{2/3} E^{-2/3} R^{-1/3},$$

where w is the load on each asperity, i.e. $w = W/(A/\lambda^2)$, and the roughness of the surface may be represented by the ratio of πR^2 to λ^2, i.e.

$$h = k_1 (W\lambda^2/A)^{2/3} (E^{-2/3} R^{-1/3}) (\pi R^2/\lambda^2).$$

Hence in (7.20)

$$\mu K_e = k' (A/W)^{1/3} (R/\lambda)^{5/3} E^{-2/3} M^{-1}$$

However, the number of cycles to failure in a fatigue situation may be written as $M = (\sigma_0/\sigma)^b$, where σ_0 is the tensile strength in a simple short term test and σ is the stress applied in the fatigue test. Therefore, equating σ with the mean stress on an asperity ($\sigma = \text{const} \times w^{1/3} E^{2/3}/R^{2/3}$) means that μK_e can be written as

$$\mu K_e = K'' \sigma_0^{-b} (W/A)^{(b-1)/3} (R/\lambda)^{(5-2b)/3} E^{(2b-2)/3}. \qquad (7.21)$$

Reznikovskii (1960) suggests that equation (7.21) is obeyed for a number of rubbers, with values of b given in table 7.2, which gives a theoretical justification for not only the increased wear of car tyres when the vehicle is heavily laden, but also when the temperature is higher and the modulus increases.

Table 7.2. *Load index b for various rubbers*

Rubber type	Styrene-butadiene	Styrene-butadiene ($+5\,^{\circ}$C)	Natural rubber	Sodium polybutadiene
b	1.9	2.2	1.75	1.4

References

Chapter 1

Adamson, A.W. (1967) *Physical Chemistry of Surfaces*, 2nd edn, Wiley-Interscience, New York.

Andreas, J.M., Hauser, E.H. & Tucker, W.B. (1938) Boundary tension by pendant drops, *J. Phys. Chem.* **42**, 1001-19.

Aveyard, R. & Haydon, D.A. (1973) *An Introduction to the Principles of Surface Chemistry*, Cambridge University Press.

Cherry, B.W., el Mudarris, S. & Holmes, C.M. (1969) Formation of the solid liquid interface, *J. Aust. Inst. Met.* **14**, 132-7.

Davis, J.K. & Bartell, F.E. (1948) Determination of the surface tension of molten materials, *Anal. Chem.* **20**, 1182-5.

Defay, R., Prigogine, I., Bellemans, A. & Everett, D.H. (1966) *Surface Tension and Adsorption*, Longmans, London.

Dettre, R.H. & Johnson, R.E., Jr (1966) The surface tensions of some molten polyethylenes, *J. Colloid Interface Sci.* **21**, 367-77.

Dettre, R.H. & Johnson, R.E., Jr (1969) Surface tensions of perfluoroalkanes and polytetrafluoroethylene, *J. Colloid Interface Sci.* **31**, 568-9.

Edwards, H. (1968) Surface tensions of liquid polyisobutylenes, *J. Appl. Polym. Sci.* **12**, 2213-24.

Fowkes, F.M. (1969) Calculation of work of adhesion by pair potential summation, *Hydrophobic Surfaces*, Academic Press, New York, pp. 151-63.

Fowkes, F.M. & Sawyer, W.M. (1952) Contact angles and boundary energies of a low energy solid, *J. Chem. Phys.* **20**, 1650.

Fowler, R.H. & Guggenheim, E.A. (1939) *Statistical Thermodynamics*, Cambridge University Press.

Gardon J.L. (1967) Variables and interpretation of some destructive cohesion and adhesion tests, in *Treatise on Adhesion and Adhesives* (Patrick, R.L., ed.) Marcel Dekker Inc., New York, pp. 269-324.

Good, R.J. (1967) Intermolecular and interatomic forces, in *Treatise on Adhesion and Adhesives* (Patrick, R.L., ed.) Marcel Dekker Inc., New York, pp. 9-68.

Greenhill, E.B. & McDonald, S.R. (1953) Surface free energy of solid paraffin wax, *Nature* **171**, 37-8.

Harkins, W.O. & Jordan, H.F. (1930) A method for the determination of surface and interfacial tension from the maximum pull on a ring, *J. Amer. Chem. Soc.* **52**, 1751-72.

Hildebrand, J.H. & Scott, R.L. (1950) *The Solubility of Non-Electrolytes*, 3rd edn, Reinhold, New York.

Kaelble, D.H. (1971) *Physical Chemistry of Adhesion*, Wiley-Interscience, New York.

McLeod, D.B. (1923) On a relation between surface tension and density, *Trans. Faraday Soc.* **19**, 38-42.

Phillips, M.C. & Riddiford, A.C. (1966) The specific free surface energy of paraffinic solids, *J. Colloid Interface Sci.* **22**, 149-57.

Porter, A.W. (1933) The calculation of surface tension from experiment - 1 - Sessile drops, *Philos. Mag.* **15**, 163-70.

Quayle, D.R. (1953) The parachors of organic compounds, *Chem. Rev.* **53**, 439-589.

Rayleigh, Lord (1890) *Scientific Papers*, vol. 3, (1887-1892) Cambridge University Press, 1902, pp. 397-425.

Roe, R.J. (1965a) Hole theory of surface tension of polymer liquids, *Proc. Natl. Acad. Sci. USA* **56**, 819-24.

Roe, R.J. (1965b) Parachor and surface tension of amorphous polymers, *J. Phys. Chem.* **69**, 2809-10.

Roe, R.J. (1968) Surface tension of polymer liquids, *J. Phys. Chem.* **72**, 2013-17.

Sakai, T. (1965) Surface tension of polyethylene melt, *Polymer* **6**, 659-61.

Schonhorn, H. (1965) Theoretical relationship between surface tension and cohesive energy density, *J. Chem. Phys.* **43**, 2041-43.

Schonhorn, H., Ryan, F.W. & Sharpe, L.H. (1966) Surface tension of molten polychlorotrifluoroethylene, *J. Polym. Sci.* **A-4**, 538-42.

Schonhorn, H. & Sharpe L.H. (1965) Surface energetics, adhesion and adhesive joints III - Surface tension of molten polyethylene, *J. Polym. Sci.* **A-3**, 569-73.

Staicopolus, D.N. (1962) The computation of surface tension and of contact angle by the sessile drop method, *J. Colloid Interface Sci.* **1**, 439-47.

Starkweather, H.W. (1965) The surface tension of polyethylene, *Soc. Plast. Eng. Trans.* **5**, 5-6.

Stewart, C.W. & Von Frankenberg, C.A. (1968) Significant structures theory of the surface tension of polyethylene, *J. Polym. Sci.* **A-2**, 1686-88.

Wu, S. (1968) Estimation of the critical surface tension for polymers from molecular constitution by a modified Hildebrand-Scott equation, *J. Phys. Chem.* **72**, 3332-34.

Wu, S. (1969) Surface and interfacial tensions of polymer melts, I, *J. Colloid Interface Sci.* **31**, 153-61.

Wu, S. (1970) Surface and interfacial tensions of polymer melts, II, *J. Phys. Chem.* **74**, 632-38.

Wu, S. (1974) Interfacial and surface tensions of polymers, *J. Macromol. Sci.* **C10**, 1-73.

Chapter 2

Brown, R.G. & Eby, R.K. (1964) Effect of crystallisation conditions and heat treatment on polyethylene: Lamellar thickness, melting temperature and density, *J. Appl. Phys.* **35**, 1156-61.

Cassie, A.B.D. & Baxter, S. (1944) Wettability of porous surfaces, *Trans. Faraday Soc.* **40**, 546-51.

Cherry, B.W. (1971) Aspects of surface chemistry and morphology, in *Plastics - Surface and Finish* (Pinner, S.H. & Simpson, W.G., eds.) Butterworths, London, pp. 217-42.

Fowkes, F.M. (1964) Dispersion force contributions to surface and interfacial tensions, contact angles and heats of immersion, in *Advances in Chemistry Series*, no. 43 (Gould, R.F., ed.) *Amer. Chem. Soc.* pp. 99-111.

Fowkes, F.M. (1967) Surface chemistry, in *Adhesion and Adhesives*, vol. 1 (Patrick, R.L., ed.) Edward Arnold, London, pp. 325-457.

Fowkes, F.M. (1969) Calculation of work of adhesion by pair potential summation, in *Hydrophobic Surfaces* (Fowkes, F.M., ed.) Academic Press, New York, pp. 151-63.

Fowkes, F.M. & Harkins, W.O. (1940) The state of monolayers adsorbed at the interface solid-aqueous solution, *J. Amer. Chem. Soc.* **62**, 3377-86.

Girifalco, L.A. & Good, R.J. (1957) A theory for the estimation of surface and interfacial energies, I - Derivation and application to interfacial tension, *J. Phys. Chem.* **61**, 904-9.

Good, R.J. (1967) Intermolecular and interatomic forces, in *Adhesion and Adhesives*, vol. 1 (Patrick, R.L., ed.) Edward Arnold, London, pp. 9-68.

Good, R.J. (1973) Comparison of contact angle interpretations, *J. Colloid Interface Sci.* **44**, 63-71.

Gornick, F. & Hoffman, J.D. (1966) Nucleation in polymers, *Ind. Eng. Chem.* **58**(2), 41-53.

Johnson, R.E. (1959) Conflicts between Gibbsian thermodynamics and recent treatments of interfacial energies in solid-liquid-vapour systems, *J. Phys. Chem.* **63**, 1655-9.

Johnson, R.E. & Dettre R.H. (1964) Contact angle hysteresis, in *Advances in Chemistry Series*, no. 43 (Gould, R.F., ed.) *Amer. Chem. Soc.* pp. 112-35.

Johnson, R.E. & Dettre, R.H. (1969) Wettability and contact angles, in *Surface and Colloid Sciences*, vol. 2 (Matijevic, E., ed.) Wiley, New York, pp. 85-153.

Lester, G.R. (1967) Contact angles on deformable solids, *Wetting*, Soc. Chem. Ind. Monograph 25, pp. 57-98.

Owens, D.K. & Wendt, R.C. (1969) Estimation of the surface free energy of polymers, *J. Appl. Poly. Sci.* 13, 1741-7.

Phillips, M.C. & Riddiford, A.C. (1966) The specific free energy of paraffinic solids, *J. Colloid Interface Sci.* 22, 149-157.

Schonhorn, H. (1965) Surface free energy of polymers, *J. Phys. Chem.* 69, 1084-5.

Wenzel, R.N (1936) Resistance of solid surfaces to wetting by water, *Ind. Eng. Chem.* 28, 988-94.

Wu, S. (1973) Polar and non-polar interactions in adhesion, *J. Adhesion* 5, 39-55.

Zisman, W.A. (1964) Relation of equilibrium contact angle to liquid and solid constitution, *Advances in Chemistry Series*, no. 43 (Gould, R.F., ed.) *Amer. Chem. Soc.* pp. 1-51.

Chapter 3

Bikerman, J.J. (1968) *The Science of Adhesive Joints*, Academic Press, New York.

Cagle, C.V. (1973) *Handbook of Adhesive Bonding*, McGraw-Hill, New York.

Cherry, B.W., el Mudarris, S. & Holmes, C.M. (1969) Formation of the solid-liquid interface, *J. Aust. Inst. Met.* 14, 132-7.

Cherry, B.W. & Holmes, C.M. (1969) Kinetics of wetting of surfaces by polymers, *J. Colloid Interface Sci.* 29, 174-6.

de Bruyne, N.A. (1939) The nature of adhesion, *The Aircraft Engineer* (supplement to Flight) 18(12), 51-4.

Glasstone, S., Laidler, K.J. & Eyring, H. (1941) The theory of rate processes, ch. 9, *Viscosity and Diffusion*, McGraw-Hill, New York.

Houwink, R. & Salomon, G., eds. (1965) *Adhesion and Adhesives*, Elsevier, Amsterdam.

Johnson, R.E. & Dettre, R.H. (1964) Contact angle hysteresis, (i) Study of an idealised rough surface and (ii) Contact angle measurements on rough surfaces, *Advances in Chemistry Series*, no. 43, (Gould, R.F., ed.) *Amer. Chem. Soc.* pp. 112-44.

Kaelble, D.H. (1971) *Physical Chemistry of Adhesion*, Wiley-Interscience, New York.

Newman, S. (1968) Kinetics of wetting of surfaces by polymers, *J. Colloid Interface Sci.* 26, 208-13.

Patrick, R.L., ed. (1969)*Treatise on Adhesion and Adhesives*, vol. 2, *Materials*, Marcel Dekker Inc., New York.

Schonhorn, H., Frisch, H.L. & Kwei, T.K. (1966) Kinetics of wetting of surfaces by polymer melts, *J. Appl. Phys.* 37, 4967-73.

Sharpe, L.H. & Schonhorn, H. (1964) Surface energetics, adhesion and adhesive joints, *Advances in Chemistry Series*, no. 43 (Gould, R.F., ed.) *Amer. Chem. Soc.* pp. 189-201.

Shields, J. (1976) *Adhesives Handbook*, Newnes-Butterworths, London.

Wake, W.C. (1965) Rubbers, in *Adhesion and Adhesives* (Houwink, R. & Salomon, G., eds.) Elsevier, Amsterdam.

Wu, S. (1973) Polar and non-polar interactions in adhesion, *J. Adhesion* 5, 39-55.

Chapter 4

Andrews, E.H. (1968) *Fracture in Polymers*, Oliver & Boyd, Edinburgh.

Andrews, E.H. & Kinloch, A.J. (1973) Mechanics of adhesive failure, *Proc. Roy. Soc.* A 332, 385-99.

Berry, J.P. (1963) Determination of fracture surface energies by the cleavage technique, *J. Appl. Phys.* 34, 62-8.

Berry, J.P. (1972) in *Fracture*, vol. 7 (Liebowitz, H., ed.) Academic Press, New York, pp. 38-93.

Bikerman, J.J. (1968) *The Science of Adhesive Joints*, Academic Press, New York.

Cherry, B.W. & Thomson, K.W. (1978) Crack stability and strain rate dependence of G_c in epoxy resin, *Int. J. Fracture* 14, R17-19.

Deryaguin, B.W. & Smilga, V.P. (1970) Electronic theory of adhesion, in *Adhesion - Fundamentals and Practice*, Applied Science Publishers, Barking, pp. 152-63.

Evans, A.G. (1972) A method for evaluating the time-dependent failure characteristics of brittle materials, *J. Mater. Sci.* 7, 1137-46.

Gent, A.N. (1972) Fracture of elastomers, in *Fracture*, vol. 7 (Liebowitz, H., ed.) Academic Press, New York, pp. 316-51.

Gent, A.N. & Kinloch, A.J. (1971) Adhesion of viscoelastic materials to rigid substrates, *J. Polym. Sci.* A-2, 9, 659-68.

Griffith, A.A. (1920), The phenomena of rupture and flow in solids, *Phil. Trans. Roy. Soc.* 221, 163-98.

Gurney, C. & Amling, H. (1970) Crack propagation in adhesive joints in *Adhesion - Fundamentals and Practice*, Applied Science Publishers, Barking pp. 211-17.

Gurney, C. & Hunt, J. (1967) Quasi-static crack propagation, *Proc. Roy. Soc.* A 299, 508-24.

Gurney, C. & Mai, Y.W. (1972) Stability of cracking, *Eng. Fracture Mech.* 4, 853-63.

Hakeem, M.I. & Phillips, M.G. (1978) Effect of environment on stability of cracking in brittle polymers, *J. Mater. Sci.* 13, 2284-7.

Irwin, G.R. (1957) Analysis of stresses and strains near the end of a crack traversing a plate, *J. Appl. Mech.* 24, 361-4.

Irwin, G.R. (1958) Fracture, in *Encyclopedia of Physics*, vol. 6 (Flügge, S., ed.) Springer Verlag, Berlin, pp. 551-90.

Kambour, R.P. (1973) Crazing and fracture in thermoplastics, *Macromol. Rev.* 7, 1-154.

Knott, J.F. (1973) *Fundamentals of Fracture Mechanics*, Butterworths, London.

Lake, G.S. & Lindley, P.B. (1966) Fatigue of rubber, in *Physical Basis of Yield and Fracture* (Stickland, A.C., ed.) Institute of Physics and Physical Society Conference Proceedings, no. 1, London, p. 177.

Mai, Y.W. (1974) Cracking stability in tapered double cantilever beam test pieces, *Int. J. Fracture* 10, 292-5.

Mai, Y.W. (1975) On the environmental fracture of polymethyl methacrylate, *J. Mater. Sci.* 10, 943-54.

Mai, Y.W. & Atkins, A.G. (1975) On the velocity dependent fracture toughness of epoxy resins, *J. Mater. Sci.* 10, 2000-3.

Marshall, G.P., Coutts, L.H., & Williams, J.G. (1974) Temperature effects in the fracture of polymethyl methacrylate, *J. Mater. Sci.* 9, 1409-19.

Moskvitin, N.I. (1969) *Physicochemical Principles of Glueing and Adhesion Processes* Israel programme for scientific translations, Jerusalem.

Paris, P.C. & Sih, G.C. (1965) Stress analysis of cracks, in *Fracture Toughness Testing and its Applications*, Amer. Soc. for Testing and Materials, STP 381, pp. 30-83.

Ripling, E.J., Mostovoy, S. & Corten, H.J. (1971) Fracture mechanics, a tool for evaluating structural adhesives, *J. Adhesion* 3, 107-23.

Rivlin, R.S. & Thomas, A.G. (1953) Rupture of rubber I, Characteristic energy for tearing, *J. Polym. Sci.* 10, 291-318.

Speake, J.H. & Curtis, G.J. (1976) Ultrasonic and stress wave emission non-destructive testing of polymeric adhesives, *Plastics and Rubber: Materials and Applications* 1, 193-204.

Timoshenko, S. (1955) *Strength of Materials - Part 1, Elementary Theory and Problems*, Van Nostrand, New York.

Westergaard, H.M. (1939) Bearing pressures and cracks, *J. Appl. Mech.* 6, 49-53.

Yamini, S. & Young, R.J. (1977) Stability of crack propagation in epoxy resins, *Polymer* 18, 1078-80.

Chapter 5

Andrews, E.H. (1974) A generalised theory of fracture mechanics, *J. Mater. Sci.* 9, 887–94.

Andrews, E.H. & Billington, E.W. (1976) Generalised fracture mechanics part 2, Materials subject to general yielding, *J. Mater. Sci.* 11, 1354–61.

Andrews, E.H. & Fukahori, Y. (1977) Generalised fracture mechanics part 3, Prediction of fracture energies in highly extensible solids, *J. Mater. Sci.* 12, 1307–19.

Andrews, E.H. & Kinloch, A.J. (1973a) Mechanics of adhesive failure I, *Proc. Roy. Soc.* A 332, 385–99.

Andrews, E.H. & Kinloch, A.J. (1973b) Mechanics of adhesive failure II, *Proc. Roy. Soc.* A 332, 401–14.

Berry, J.P. (1963) Determination of fracture surface energies by the cleavage technique, *J. Appl. Phys.* 34, 62–8.

Berry, J.P. (1972) Fracture of polymeric glasses, in *Fracture*, vol. 7 (Liebowitz, H., ed.) Academic Press, New York, pp. 38–93.

Brown, H.R. & Ward, I.M. (1973) Craze shape and fracture in poly(methylmethacrylate), *Polymer* 14, 469–75.

Cherry, B.W. & Harrison, N.L. (1970) The optimum profile for a lap joint, *J. Adhesion*, 2, 125–8.

Cherry, B.W. & Holmes, C.M. (1971) Mechanism of failure of adhesive joints, in *Aspects of Adhesion*, vol. 6, (Alner, D.J., ed.) University of London Press, 80–95.

Cherry, B.W. & Thomson, K.W. (1977) Water induced fracture of epoxy-aluminium joints, in *Fracture Mechanics and Technology*, vol. 2 (Sih, G.C., ed.) Sijthoff & Noordhoff, Netherlands, 723–37.

Dann, J.R. (1970) Forces involved in the adhesive process, *J. Colloid Interface Sci.* 32, 302–31.

de Bruyne, N.A. (1944) The strength of glued joints, *Aircraft Eng.* 16, 115–8.

Dugdale, D.S. (1960) Yielding of steel sheets containing slits, *J. Mech. Phys. Sol.* 8, 100–4.

Fowkes, F.M. (1967) Surface chemistry, in *Treatise on Adhesion and Adhesives*, vol. 1, (Patrick, R.L., ed.) Marcel Dekker Inc., New York, pp. 325–449.

Gent, A.N. (1972) Fracture of elastomers, in *Fracture*, vol. 7 (Liebowitz, H., ed.) Academic Press, New York, 315–50.

Gent, A.N. & Lindley, P.B. (1959) Internal rupture of bonded rubber cylinders in tension, *Proc. Roy. Soc.* A 249, 195–205.

Gent, A.N. & Schulz, J. (1972) Effect of wetting liquids on the strength of adhesion of viscoelastic materials, *J. Adhesion* 3, 281–94.

Gledhill, R.A. & Kinloch, A.J. (1974) Environmental failure of structural adhesive joints, *J. Adhesion* 6, 315–30.

Hang, P.T.T. (1975) Shrinkage stresses in adhesive joints, M. Eng. Sci. thesis, Monash University.

Kaelble, D.H. (1970) Dispersion/polar surface tension properties of organic solids, *J. Adhesion*, 2, 66–81.

Knott, J.F. (1973) *Fundamentals of fracture mechanics*, Butterworths, London.

Lake, G.J. & Lindley, P.B. (1965) The mechanical fatigue limit for rubber, *J. Appl. Polym. Sci.* 9, 1233–51.

Lake, G.J. & Thomas, A.G. (1967) The strength of highly elastic materials, *Proc. Roy. Soc.* A 300, 109–19.

Malyshev, B.M. & Salganik, R.L. (1965) The strength of adhesive joints using the theory of cracks, *Int. J. Fract. Mech.* 1, 114–28.

Mostovoy, S., Ripling, E.J. & Bersch, C.F. (1971) Fracture toughness of adhesive joints, *J. Adhesion* 3, 125–44.

Mullins, L. (1959) Role of hysteresis in the tearing of rubber, *Trans. Inst. Rubber Industry* 35, 213–22.

Rice, J.R. & Sih, G.C. (1965) Plane problems of cracks in dissimilar media, *J. Appl. Mech* **32** 418–23.

Ripling, E.J., Mostovoy, S. & Bersch, C.F. (1971) Stress corrosion cracking of adhesive joints, *J. Adhesion* **3**, 145–63.

Rivlin, R.S. & Thomas, A.G. (1953) Rupture of rubber – I Characteristic energy for tearing, *J. Polym. Sci.* **10**, 291–318.

Sneddon, I.N. (1961) The distribution of stress in adhesive joints, in *Adhesion* (Eley, D. D., ed.) Oxford University Press, pp. 207–53.

Treloar, L.R.G. (1958) *The Physics of Rubber Elasticity*, Oxford University Press.

Vincent, P.I. & Gotham, K.V. (1966) Effect of crack propagation velocity on the fracture surface energy of poly(methylmethacrylate), *Nature* **210**, 1254.

Wang, T.T., Kwei, T.K. & Zupko, H.M. (1970) Tensile strength of butt-jointed epoxy–aluminium plates, *Int. J. Fract. Mech.* **6**, 127–37.

Williams, M.L., Landel, R.F. & Ferry, J.D. (1955) The temperature dependence of relaxation mechanisms in amorphous polymers and other glass forming liquids, *J. Amer. Chem. Soc.* **77**, 3701–7.

Williams, J.G. & Marshall, G.P. (1975) Environmental crack and craze growth in polymers. *Proc. Roy. Soc.* A **342**, 55–77.

Chapter 6

Aird, P.J. & Cherry, B.W. (1978) Frictional properties of branched polyethylene II. The normal force–shear-force relationship, *Wear* **51**, 147–55.

Allen, A.J.G. (1958) Plastics as solid lubricants and bearings, *Lubr. Eng.* **14**, 211–15.

Archard, J.F. (1957) Elastic deformation and the laws of friction, *Proc. Roy. Soc.* A **243**, 190–205.

Bowden, F.P. & Tabor, D. (1950) *The Friction and Lubrication of Solids*, part 1, Oxford University Press.

Bowden, F.P. & Tabor, D. (1964) *The Friction and Lubrication of Solids*, part 2, Oxford University Press.

Bowden, F.P. & Tabor, D. (1966) Friction, lubrication and wear: a survey of the work done in the last decade, *Brit. J. Appl. Phys.* **17**, 1521–44.

Bridgeman, P.W. (1935) Effects of high shearing stress combined with high hydrostatic pressure, *Phys. Rev.* **48**, 825–47.

Briscoe, B.J. & Tabor, D. (1975) The effect of pressure on the frictional properties of polymers, *Wear* **34**, 29–38.

Cottrell, A.H. (1964) *The Mechanical Properties of Matter*, Wiley, New York.

Gent, A.N. (1958) On the relationship between indentation hardness and Young's modulus, *Trans. Inst. Rubber Ind.* **34**, 46–57.

Glasstone, S., Laidler, K.J. & Eyring, H. (1941) *The Theory of Rate Processes*, McGraw-Hill, New York.

Greenwood, J.A. (1967) The area of contact between rough surfaces and flats, *J. Lub. Technol.* **1**, 81–91.

Greenwood, J.A. & Williamson, J.B.P. (1966) Contact of nominally flat surfaces, *Proc. Roy. Soc.* A **295**, 300–19.

Grosch, K.A. (1963) The relation between the friction and visco-elastic properties of rubber, *Proc. Roy. Soc.* A **274**, 21–39.

Hatfield, M.R. & Rathman, G.B. (1956) Application of the absolute rate theory to adhesion, *J. Phys. Chem.* **60**, 957–61.

Hertz, H. (1882) Über die Berührung fester elasticher Korper, *J. Reine und Angewandte Mathematik* **92**, 156–71.

Hill, R. (1960) *The Mathematical Theory of Plasticity*, Oxford University Press.

Johnson, K.L., Kendall, K. & Roberts, A.D. (1971) Surface energy and the contact of elastic solids, *Proc. Roy. Soc.* A **324**, 301–13.

Lancaster, J.K. (1972) Friction and wear, in *Polymer Science*, vol. 2 (Jenkins, A.D., ed.) North-Holland, Amsterdam, pp. 960–1046.

Lee, L.H. (1974) Effect of surface energies on polymer friction and wear, in *Advances in Polymer Friction and Wear* vol. 1 (Lee, L. H., ed.) Plenum Press, New York.

Ludema, K.C. & Tabor, D. (1966) The friction and viscoelastic properties of polymeric solids, *Wear* 9, 329–48.

Lur'e, A.I. (1964) *Three-Dimensional Problems on the Theory of Elasticity*, Interscience, London.

Makinson, K.R. & Tabor, D. (1964) The friction and transfer of polytetrafluoroethylene, *Proc. Roy. Soc.* A 281, 49–61.

McLaren, K.G. & Tabor, D. (1963) Friction of polymers: influence of speed and temperature, *Nature* 197, 856–8.

Pascoe, M.W. & Tabor, D. (1956) The friction and deformation of polymers, *Proc. Roy. Soc.* A 235, 210–24.

Pooley, C.M. & Tabor, D. (1972) Friction and molecular structure: the behaviour of some thermoplastics, *Proc. Roy. Soc.* A 329, 251–74.

Savkoor, A.R. (1974) Adhesion and deformation friction of polymers on hard solids, in *Advances in Polymer Friction and Wear*, vol. 1 (Lee, L.H., ed.) Plenum Press, New York, pp. 69–122.

Schallamach, A. (1963) A theory of dynamic friction, *Wear* 6, 375–82.

Smith T.L. (1958) Dependence of the ultimate properties of a GRS rubber on strain rate and temperature, *J. Polym. Sci.* 32, 99–113.

Sternstein, S.S. & Ongchin, L. (1969) Yield criterion for plastic deformation of glassy high polymers in general stress fields, *Polymer preprints*, Sept. 1969, *Amer. Chem. Soc.* pp. 1117–24.

Tanaka, K. (1961) Friction and deformation of polymers, *J. Phys. Soc. Japan* 16, 2003–16.

Timoshenko, S.P. & Goodier, J.N. (1970) *Theory of Elasticity,* 3rd edn, McGraw-Hill, New York.

Towle, L.C. (1974) Shear strength and polymer friction, in *Advances in Polymer Friction and Wear* (Lee, L.H., ed.) Plenum Press, New York, pp. 174–90.

Williams, J. G. (1973) *Stress Analysis of Polymers*, Longmans, London.

Chapter 7

Barquins, M. & Courtel, R. (1975) Rubber friction and the rheology of viscoelastic contact, *Wear* 32, 133–50.

Barquins, M., Courtel, R. & Thirion, P. (1974) Sur l'observation des ondes de Schallamach et leur rôle dans le frottement du caoutchouc, *Wear* 27, 147–50.

Bowden, F.P. & Tabor, D. (1964) *The Friction and Lubrication of Solids*, Oxford University Press.

Bowden, F.P. & Tabor, D. (1966) Friction, lubrication and wear: a survey of work during the last decade, *British J. Appl. Phys.* 17, 1521–44.

Briggs, G.A.D. & Briscoe, B.J. (1975) The dissipation of energy in the friction of rubber, *Wear* 35, 357–64.

Champ, D.H., Southern, E. & Thomas, A.G. (1974) Fracture mechanics applied to rubber abrasion, in *Advances in Polymer Friction and Wear* (Lee, L.H., ed.) Plenum Press, New York, pp. 133–42.

Flom, D.G. & Bueche, A.M. (1959) Rolling friction of a hard cylinder over a viscoelastic material, *J. Appl. Phys.* 30, 1725–30.

Greenwood, J.A., Minshall, H. & Tabor, D. (1961) Hysteresis losses in rolling and sliding friction, *Proc. Roy. Soc.* A 259, 480–507.

Greenwood, J. A. & Tabor, D. (1958) The friction of hard sliders on lubricated rubbers: the importance of deformation losses, *Proc. Phys. Soc.* 71, 989–98.

Grosch, K.A. (1963) The relation between the friction and visco-elastic properties of rubber, *Proc. Roy. Soc.* A 274, 21–39.

Johnson, K.L., Kendall, K. & Roberts, A.D. (1971) Surface energy and the contact of elastic solids, *Proc. Roy. Soc.* A 324, 301–13.

Kendall, K. (1974) Kinetics of contact between smooth solids, *J. Adhesion* 7, 55–72.

Kendall, K. (1975) Rolling friction and adhesion between smooth solids, *Wear* 33, 351–58.

Lancaster, J.K. (1973) Basic mechanisms of friction and wear of polymers, *Plast. Polym.* 41, 297–306.

Moore, D.F. (1972a) On the decrease in contact area for spheres and cylinders rolling on a viscoelastic plane, *Wear* 21, 179–94.

Moore, D.F. (1972b) *The Friction and Lubrication of Elastomers*, Pergamon Press, Oxford.

Moore, D.F. & Geyer, W. (1972) A review of adhesion theories for elastomers, *Wear* 22, 113–41.

Reznikovskii, M.M. (1960) Relation between the abrasion resistance and other mechanical properties of rubber, *Sov. Rubber Technol.* 19, no. 9, 32–7. (Included in *Abrasion of Rubber* (James, D.T., ed.). McLaren and Sons, London, 1967, pp. 119–26.)

Schallamach, A. (1958) Friction and abrasion of rubber, *Wear* 1, 384–417.

Schallamach, A. (1971) How does rubber slide? *Wear* 17, 301–12.

Thomas, A.G. (1958) Rupture of rubber V. Cut growth in natural rubber vulcanisates, *J. Polym. Sci.* 31, 467–80.

Timoshenko, S.P. & Goodier, J.N. (1970) *Theory of Elasticity*, 3rd edn, McGraw-Hill, New York.

Author Index

Subject Index